COSMIC GAS DYNAMICS

We want to thank the following people in connection with the preparation of the manuscript for publication: Elaine Dodge, A. Malinda Sanborn, and Shirley Iverson of the Aerospace Engineering Sciences Staff, and Dr. Andrew Skumanich, Professor Adjoint of the Department of Physics and Astrophysics, University of Colorado, and Research Staff Member of the High Altitude Observatory, Boulder.

Professor Biermann wishes to acknowledge his indebtedness to A. Busemann, K. Jockers, F. Meyer, D. Pfirsch, and H. U. Schmidt for suggested corrections and additions to these notes, to W. Huebner for reading carefully the entire text, and to Mrs. G. Wessling, for the painstaking technical assistance in completing the manuscript.

These works are based on lectures we gave during the summer of 1968 under the Aersospace Engineering Sciences Aerospace Frontiers Program. This Program began in 1963 and the following subjects have been covered:

- 1963: Lectures on phase transition, Professor George E. Uhlenbeck, Rockefeller University
- 1964: The problems of fire control, paper-pulp processing, and high-temperature gas flow in pipes, Professor Howard Emmons, Harvard University
- 1965: Stochastic processes, Professor Uhlenbeck
- 1966: The Boltzmann equation, Professor Uhlenbeck
- 1967: Selected topics in fluid dynamics, Professor Paul Robert Owen, Imperial College, London: Advanced thermodynamics, Professor Uhlenbeck
- 1968: Quantum fluid mechanics, Professor Uhlenbeck, with Professor Michael Sanders, University of Michigan; Dnyamics of interplanetary material, Professor Ludwig Biermann, Max Planck Institute, Munich; Cosmic gas dynamics, Professor Evry Schatzman, University of Paris
- 1969: Introduction to the general theory of relativity, Professor Uhlenbeck; Cosmological observations and experiments in general relativity, Professor Biermann
- 1970: Quantum statistical mechanics, Professor Uhlenbeck; Pulsars and other late stages of stellar evolution, Professor Biermann
- 1971: Topics in stochastic processes, Professor Uhlenbeck; Topics in cometary fluid mechanics, Professor Biermann
- 1972: Kinetic theory of gases, Professor Uhlenbeck; Cosmic plasma dynamics, Professor Biermann; Cosmology, including antimatter, Professor Schatzman
- 1973: Principles of thermodynamics and statistical mechanics, Professor Uhlenbeck; Astrology and astronomy through the ages, Professor Engebert Schücking, New York University

Notes on these other subjects may be made available in the future.

MAHINDER S. UBEROI, *Editor*

Department of Aerospace Engineering Sciences
University of Colorado, Boulder

COSMIC GAS DYNAMICS

Part I: Cosmic Gas Dynamics

EVRY SCHATZMAN

Institute of Astrophysics
University of Paris

Part II: Dynamics of Interplanetary Material

LUDWIG BIERMANN

Max Planck Institute for Astrophysics
Munich

Edited by

MAHINDER S. UBEROI

A WILEY-INTERSCIENCE PUBLICATION

JOHN WILEY & SONS

New York • London • Sydney • Toronto

Permission granted by the University of Chicago Press for use of the following tables and figures:

a. From *The Astrophysical Journal, Copyright . . . by the University of Chicago*: Tables 7—Allan Sandage, Vol. 125 (1957); 9 & 10—D. E. Osterbrock, Vol. 134 (1961); 13—Robert P. Kraft, Vol. 150 (1967). Figures 2—Helmut A. Apt and James H. Hunter, Jr., Vol. 136 (1962); 3—R. P. Kraft, Vol. 142 (1965); 4—Allan Sandage, Vol. 125 (1957); 7b—Dimitri Mihalas, Vol. 141 (1965); 11—Robert W. Noyes and R. B. Leighton, Vol. 138 (1963); 16—E. N. Parker, Vol. 122 (1955); 18—Robert E. Danielson, Vol. 134 (1961); 21a—I. J. Danziger and P. S. Conti, Vol. 146 (1966); 21b—A. E. Merchant, Vol. 147 (1967); 21c—P. S. Conti and I. J. Danziger, Vol. 146 (1966).
b. From *Basic Astronomical Data III*, edited by K. Aa. Strand, © *by the University of Chicago. Published 1963. Composed and printed by The University of Chicago Press, Chicago, Illinois, U.S.A.*: Tables 1—Philip C. Keenan, 4–6—Bengt Strömgren; Figure 1—A. Blaauw.

For full references and credits for all figures and tables used, see appropriate "List . . . "

Library of Congress Cataloging in Publication Data:

Main entry under title:
Cosmic gas dynamics.

"A Wiley-Interscience publication."
"Based on lectures given by the authors during the summer of 1968 under the Aerospace Engineering Sciences Aerospace Frontiers Program."
1. Gas dynamics. 2. Interstellar matter.
3. Cosmic physics. I. Schatzman, Evry L. Cosmic gas dynamics. 1973. II. Biermann, Ludwig, 1907– Dynamics of interplanetary material. 1973.

QC809.F5C67 533′.2 73–16025
ISBN 0–471–75720–9

Printed in the United States of America

10 9 8 7 6 5 4 3 2 1

PREFACE

For quite some years it has been recognized that motions of gases in the atmosphere and in the corona of stars and the effect of magnetic field on these motions are important in understanding observational data and in developing adequate theories for stellar structure and evolution.

Evry Schatzman and Ludwig Biermann have played a prominent part in pointing out the importance of gasdynamics in astrophysics. In addition to astrophysics, researchers from the areas of physics, applied mathematics, fluid mechanics and plasma dynamics have contributed to the field, and the book should be of interest to students and workers in these fields.

Although significant contributions have been made, much work needs to be done, which will no doubt be based on the pioneering work of these two authors.

The first part of the book is by Evry Schatzman. It begins with an astrophysical introduction on stellar classification and stellar evolution which should be helpful to non-astrophysicists. The second chapter is on the structure of the outer layers of a star and includes convection, turbulent velocities, waves and shock waves. The third chapter deals with flare stars, solar flares, the mechanism of the flash phase and solar cosmic rays. The fourth chapter deals with nuclear reactions and accelerating mechanisms. Finally, the role of electromagnetic activity in stellar evolution is considered, with emphasis on the energy balance.

The second part of the book, by Ludwig Biermann, deals with the dynamics of interplanetary material, and is a natural extension of the first part, which deals with stellar atmospheres with emphasis on solar atmosphere and surface activities. The subject of solar corona and solar wind is discussed from a hydrodynamic point of view. These topics are then discussed from the point of view of kinetic theory. Large- and small-scale phenomena related to magnetic fields and solar winds are discussed. The termination of the solar wind by the outer shock front is the subject of the final chapter in this part.

Mahinder S. Uberoi

TABLE OF CONTENTS

TABLE OF CONTENTS--
COSMIC GAS DYNAMICS, continued

TABLE OF CONTENTS--

DYNAMICS OF INTERPLANETARY MATERIAL, continued

TABLE OF CONTENTS, continued

LIST OF TABLES AND THEIR SOURCES

COSMIC GAS DYNAMICS

LIST OF TABLES AND THEIR SOURCES--

COSMIC GAS DYNAMICS, continued

DYNAMICS OF INTERPLANETARY MATERIAL

LIST OF FIGURES AND THEIR SOURCES

COSMIC GAS DYNAMICS

LIST OF FIGURES AND THEIR SOURCES

COSMIC GAS DYNAMICS, continued

LIST OF FIGURES AND THEIR SOURCES

COSMIC GAS DYNAMICS, continued

COSMIC GAS DYNAMICS

COSMIC GAS DYNAMICS

by

E. Schatzman

Institute of Astrophysics, University of Paris

Lectures Presented During the Summer Session, 1968

(Additions Made to Text, 1970)

in the

DEPARTMENT OF AEROSPACE ENGINEERING SCIENCES

University of Colorado

Boulder, Colorado

I. Astrophysical Introduction

A. General Considerations. The problem of the origin and evolution of the stars is certainly one of the main problems of astrophysics. Considerable progress has been achieved during the last 30 years; however, there remain a number of unsolved mysteries. More information derived from the observations and a better understanding of the physical processes are both necessary to solve these mysteries. It is the purpose of these lectures to show which path to follow in order to reach the solution of the problems still remaining.

A certain number of basic astrophysical considerations have to be given first, in order to make clear on which facts we may rely. We shall introduce theoretical considerations which are generally accepted. We shall then see that a number of physical factors which are usually considered as of little importance for stellar evolution have to be studied. We shall try to show how these physical factors give, at least qualitatively, the possibility of understanding many difficulties raised by the observations.

B. Stellar Classification. It is certainly not necessary to give here an extended description of the classification of stars. As is well known, an important classification is a two-dimension classification based on the spectral type and on the luminosity of the stars. In practice, instead of the luminosity, the absolute visual magnitude is used; and the resulting diagram, showing the place of the stars, is called the Hertzsprung-Russell diagram.*

1. Magnitudes. Using a photoelectric receiver and a coloured filter, it is possible to measure the flux per unit surface.

$$m_V = -2.5 \log E_V + \text{constant}.$$

The choice of the constant results from a number of conventions made for historical reasons and consists in choosing the value of m_V for

---

*The Hertzsprung-Russell diagram is described in all elementary books on astrophysics. See, for example, O. Struve (1959).

a certain number of well-known stars.

E_V is related to the luminosity L_V of the star by the relation

$$E_V = \frac{L_V}{4\pi d^2}$$

where d is the distance of the star. This gives

$$m_V = -2.5 \log L_V + 5 \log d + \text{constant}$$

For a special choice of d, one obtains the absolute magnitude,

$$M_V = -2.5 \log L_V + 5 \log d_0 + \text{constant}$$

If one takes the parsec as a unit, d_0 is taken to be 10 parsecs, and one obtains

$$m_V = M_V - 5 + 5 \log d$$

Once the distance is known, the absolute magnitude can be derived from the apparent magnitude.

In the scale of the absolute magnitude, the absolute visual magnitude of the sun is

$$M_{V_\odot} = 4.63$$

2. Spectral Types. Only a quantitative analysis of a stellar spectrum can give reliable information on the physical properties of the atmosphere and on the chemical composition. However, a large number of results have been derived from spectral classifications. As explained by Keenan (1963), spectral classification is the direct estimation of the physical characteristics of stars by comparison of their spectral features. The method consists essentially in placing the spectrum of a given star in a sequence of spectra of stars, in which the physical conditions are approximately known.

To classify the spectra, it is necessary to have a sufficiently large number of objects, to make a choice of criteria, to make the ordering of the spectra according to the change of intensity of the lines, and to make the calibration of the types in terms of the physical characteristics of the stars.

The choice of the criteria is not arbitrary, and there is presently an attempt to make such a choice that physical variables can be separated. First of all, one can expect to find features sensitive to the temperature and to the pressure. As the pressure is related to the surface gravity, it is possible to separate stars having the same surface temperature according to their radii. This is equivalent to finding the luminosity of the stars, and some criteria give, at least approximately, the absolute magnitude of the stars.

Next, it is possible to interpret other features as due to peculiarities in the chemical abundance, to the magnetic field, or to Doppler broadening of the lines by hydrodynamic motion.

As we shall use extensively the following references to the spectral types, it is necessary to give briefly the criteria which are used by Morgan Keenan Kellmann (1943) (See also the table p. 253 in Pecker, Schatzman)

Type O: He II 4200 and 4542 to He I 4471
He II 4200 and 4542 to Hγ + Hδ

Type B: He I lines maximum at B2
He I 4121-4144 to Si II 4128-4130 (B2 to B8)

Type A: several parallel sequences of peculiar A stars and metallic line stars
K of Ca II to Hδ

Type F: Hδ to Fe 4045
Hγ to Ca I 4226

Types G: and K: Fe 4144 to Hδ
G band of CH 4300 to Mn I 4030
Ca I 4226 to Fe 4325

To these spectral types correspond effective temperatures, determined for certain stars from a quantitative analysis of the spectra. As we shall use later the reference to these effective temperatures, it is necessary to give here the provisional scale for MK types (Keenan, 1963) (Table I).

Table I*

Provisional Scale of effective temperatures for MK types

Types	T_e (°K)					
B0	27,000					
B1	23,000					
B2	20,000					
B3	18,000					
B5	16,000					
B6.5	14,000					
B8	12,500					
B9	11,200					
A0	10,400					
A1	9700					
A2	9100					9170
A3	8500					
A5	8200					
A7	7600					
F0	7200					
	Main-Sequence Subgiants		Giants		Supergiants	
	V	IV	III	II	Ib	Ia
F2	6900	6830	6800	6700	6600	....
F5	6700	6600	6500	6350	6200	....
F6	6500	6370	6250	6020	5800	....
F8	6200	6050	5900	5720	5450	....
G0	6000	5720	5500	5350	5050	....
G2	5740	5420	5100	4950	4750	....
G5	5520	5150	4800	4650	4500	....
G8	5320	4950	4600	4450	4300	....
K0	5120	4750	4400	4350	4100	....
K1	4920	4550	4150	4000	3850	....
K2	4760	....	3970	3860	3750	....
K3	4600	....	3820	3720	3600	....
K5	4350	....	3700	3600	3500	....
K6	4000	....	....	....	....	....
M0	3750	....	3500	3400	3300	....
M1	3600	....	3300	3150:	3050	....
M2	3350:	....	3100	2050:	....	....
M3	3100:	....	2900:	....	....	....
M4	....	....	2700:	....	....	....

The scale of absolute magnitudes has only a relative value, as there is certainly a continuous sequence of stars from the dwarfs to the supergiants, as shown by the study of stellar clusters. Fig. 1 gives the visual absolute magnitudes of luminosity classes, in the **MKK system.**

3. Peculiar A and Metallic line Stars. The existence, near the spectral type A of spectral peculiarities, has been known for a long time. Peculiar A stars are noted Ap in the HD catalogue.

* Table in Keenan (1963).

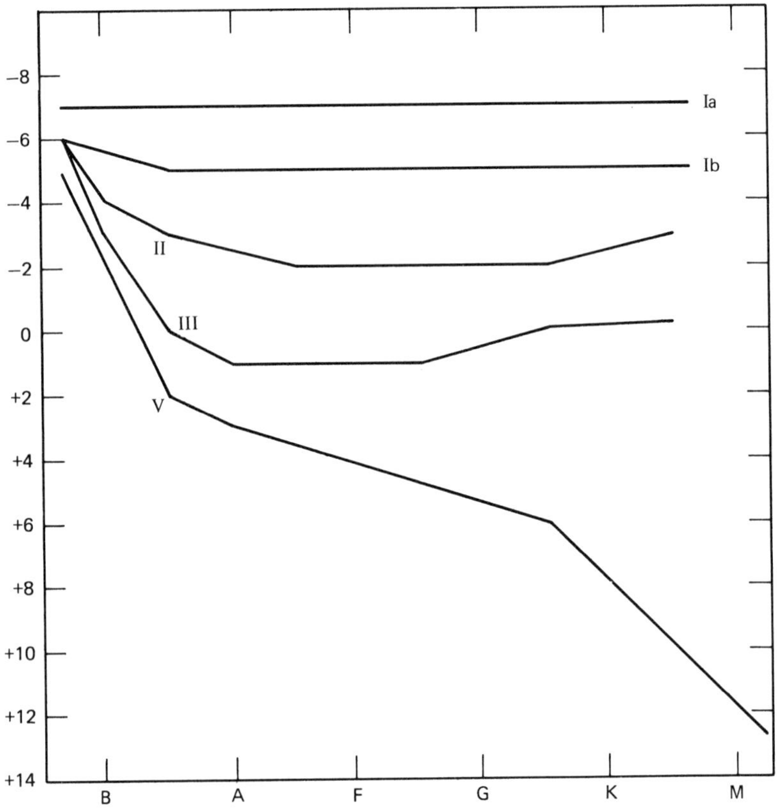

Fig. 1
MKK Luminosity Classes

Bertaud (1959, 1960) has given a catalogue and bibliography of all the A stars with peculiarities.

Due to the great variety of spectral anomalies it is very difficult to give a clear cut identification of these stars. We shall refer mainly here to the review article of Sargent (1964), whereas we shall later on use other sources of information.

Peculiar A stars have colors and hydrogen line strength similar to normal A and late B stars. However, they do not fit into the two-dimensional classification scheme found by Morgan (1943). It is possible to classify them according to the predominant peculiar element. We then obtain the following table (Table II). Essentially, these stars are more blue than it seems from their HD spectral types (higher temperature). But the classification is certainly not as simple as it seems to result from Table II. Sargent gives a number of examples of difficulties. Some stars have both strong silicon and strontium lines, which occur at both ends of the sequence; 3 Cen A and L Can have normally strong lines of phosphorous II.

Table II

Mean Colors and Spectral Types (Jashek classification)

Peculiarity	Mean HD Spectral Type	Mean B-V	Mean Spectral Type from Color
λ4200Si	A0	-0.13	B 6 V
Mn	B8-9	-0.09	B 8 V
Si	A0	-0.08	B 8-9 V
Si Eu Cr	A0	-0.07	B 9 V
Eu Cr	A1	+0.01	A 0 V
Eu Cr Sr	A2	+0.09	A 3 V
Sr	A3	+0.13	A 4 V

It seems that Ap stars in clusters or moving groups fall near the turnoff point and on the evolved main sequence of the color-magnitude diagram of the particular cluster or group. There might be a correlation between the kind of peculiarity and the age (Table III). A preliminary picture is that a few A stars temporarily develop the characteristics of an Ap star as they approach the ends of their lives on the main sequence.

Table III

Relation between Peculiarity and Age

Peculiarity	Age	Spectral type at turnoff point
λ4200 Si	$2.10^{7}-10^{8}$	B 8
Eu Cr Sr	$1-6.10^{8}$	B 9
metallic line stars	$3.10^{8}-10.10^{8}$	A 2 F 9

Peculiar A stars appear to rotate less rapidly than normal main sequence A stars. Thus, Slettebak (1954) found a mean equatorial velocity $<v_e>$ of 177 km/sec for 87 main sequence stars between B8 and A2, and a $<v_e>$ of 52 km/sec^{-1} for 16 Ap stars. It is quite possible that unusual features are invisible in fast rotating stars, and that Ap stars are rapidly rotating stars seen pole-in. It is also possible, according to Sargent (1964) that Ap stars differ in their rotational properties (or, at any rate, line sharpness) according to the kind of peculiarity--for example, there are Mn stars having ultra-sharp lines, and others having relatively broad lines, whereas there are no 4200 Si stars with ultra-sharp lines.

Ap stars are found in normal proportion in visual binaries, whereas there is definitely a smaller fraction of spectroscopic binaries among the Ap stars. Peculiar A stars all probably have a strong magnetic field. A strong magnetic field has been found by Babcock for 70 stars. The magnetic field is generally variable (α≡periodic with fluctuation, β≡fluctuating with reversal of polarity, γ≡same polarity with irregular variations). The most extreme case is HD215441 with a magnetic field varying from 12000 to 34000 gauss. The α variables are also spectrum variables. The best example is α2 CV_n.

Quantitative analysis of peculiar A stars show a number of remarkable abundance peculiarities. These will be discussed with the problem of nuclear reactions at the surface of the stars.

Metallic line stars can be characterized spectroscopically in the following way: (a) the K line of Ca II is weak, sharp and deep and corresponds to a spectral type in the range A1 to A5, (b) the hydrogen lines lead to a type in the range A8 to F0, (c) the metallic lines are strong, particularly those of the ionized metals and of the rare earths; they are also relatively sharp.

Am stars rotate slowly, with $<v \text{ Sin } i> = 40 \text{ km sec}^{-1}$. It seems that all Am stars, or at least a high percentage, are members of spectroscopic binary systems.

Quantitative analysis of the Am stars definitely shows abundance peculiarities, but they are not as extreme as those of the Ap stars. The characteristics of the spectrum seem to be due to an abnormally high value of the micro-turbulent velocity and a low spectroscopic surface gravity. To summarize, it seems that Am stars have (a) a normal abundance of Mg, Al, Ti, V, Mn, Fe, and Co; (b) a deficiency of Ca and Sc by a factor around four; (c) an overabundance of Ni and Zn by about the same factor.

4. Stellar Rotation. The spectral lines of a star are broadened by rotation. This gives the possibility of estimating the quantity $<v_e \text{ Sin } i>$ for any star. The most remarkable fact is probably the distribution of $<v_e \text{ Sin } i>$ as a function of spectral type (Fig. 2). For main sequence stars, the average $<v_e \text{ Sin } i>$ drops very quickly at spectral type F2.

Especially interesting in that respect are the results of Kraft (1965) concerning the Hyades and Coma clusters, which show the same drop in rotational velocity (Fig. 3) and also show the Ap stars as slow rotators. Problems of interpretation of the distribution of rotational velocities as a function of spectral type will be discussed at length later.

5. Color Indices and Spectral Types. A system of wide band photometry, the UBV system has been used for quite a long time.

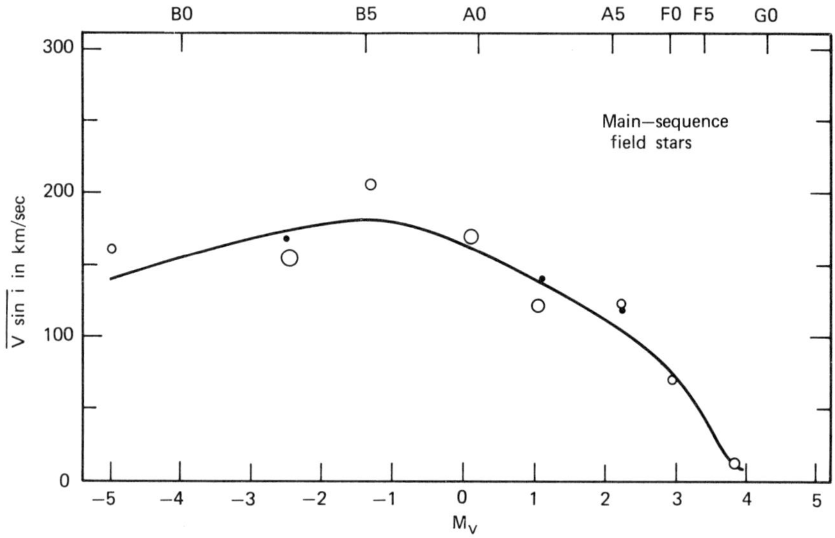

FIG. 2–Slettebak and Howard's and Herbig and Spalding's measures of mean rotational velocities of O–GO main sequence stars are plotted against absolute visual magnitude (Arp 1958) or spectral types. The sizes of the circles indicate the number of stars incorporated in the means, namely, 15–65 stars. The dots near the Al–A3 and A4–A7 means represent the means after removal of the Ursa Major stream stars from this sample of field stars, indicating that the low rotational velocities of A0–A3 stars and high rotational velocities for A4–A7 stars in the Ursa Major streams are the cause of the apparent dip at A2. Similarly, the removal of the 21 Cassiopeia-Taurus group stars from the sample of 65 B1–B3 field stars partially removes the apparent dip at B2 (see *dot* at B2).

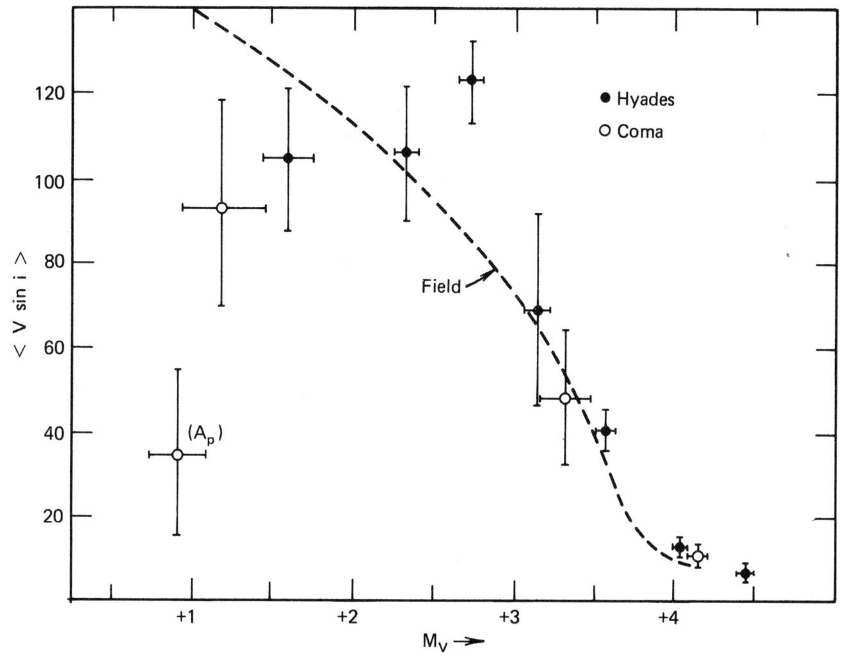

FIG. 3—The quantity ($V \sin i$) as a function of M_V in the Hyades, Coma, and in the field. The field curve is taken from the work of Abt and Hunter (1962). Horizontal and vertical lines have lengths equal to twice the standard deviation. All binaries have been omitted except those known to have $P > 10$ days; all Am stars have been omitted.

The correspondence between MK spectral types and indices B-V, U-B is given in Table IV.

Table IV

Main Sequence Stars

Colors and MK Spectral Types

	B-V	U-B		B-V	U-B		B-V	U-B
B0	-0.32	-1.13	A3	+0.09	+0.07	G5	+0.68	0.21
1	-28	-1.00	5	+0.15	+0.09	8	0.70	0.24
2	-24	-0.86	7	+0.19	+0.08	K0	0.82	0.48
3	-20	-0.71	F0	+0.30	+0.02	1	0.86	0.54
5	-16	-0.56	F2	+0.37	+0.00	3	1.01	0.89
7	-13	-0.47	F5	+0.44	+0.00	K5	1.18	1.12
8	-09	-0.29	6	+0.47	-0.02	7	1.37	1.26
9	-05	-0.16	8	+0.53	+0.02	M1	1.48	1.21
A0	-00	0.00	G0	+0.60	+0.06	3	1.49	1.10
1	+05	0.05	2	+0.64	+0.16	5	1.69	1.24

The large band photometry gives access to a larger number of stars. However, more information can be gained on the spectral properties by using filters of smaller band width, and better placed in the spectrum. Strömgren (1963) uses the following (Table V). These filters have been used for studying 1200 stars. These indices are very useful in the interval A2-G0.

Table V

Central Wavelength and Band Width used by Strömgren

Band	λ	$\Delta\lambda$	Band	λ	$\Delta\lambda$
u	3500	300	b	4670	180
v	4100	190	y	5470	230

The index b-y (where b and y are magnitudes) is close to B-V but is less influenced by the presence of the spectral lines; the color difference $c_1=(u-v)-(v-b)$ is a measure of the Balmer discontinuity; the index b-y is an indicator of effective temperatures, not much affected by the metal content, but the color index difference $m_1=(v-b)-(b-y)$ is a good indicator of metal content. Strömgren (1963) using Wallerstein's results gives the following relations:

$$\left[\frac{Fe}{H}\right] = \log\left(\frac{\text{abundance Fe}}{\text{abundance H}}\right)_{\text{star}} - \log\left(\frac{\text{abundance Fe}}{\text{abundance H}}\right)_{\text{sun}}$$

$$\left[\frac{Fe}{H}\right] = 0.3 - 12\ \Delta m_1$$

where Δm_1 is the difference between $(m_1)_{star}$ and $(m_1)_{Hyades}$. The quantity $(m_1)_{Hyades}$ is derived from the calibration of the stars of the Hyades.

With use of these indexes, Strömgren shows clearly the existence of differences in the influence of the metal lines on the spectrum. Let us consider first the calibration in spectral types for the sequence (Table VI).

Table VI

The standard relation between color index differences m_1, color index (b-y) (effective temperature) for zero age m.s.

b-y	m_1	B-V	Sp Class
+0.10	0.210	+0.17	A6
15	201	24	A8
20	185	31	F0
25	171	38	F2
30	171	46	F6
35	189	54	F8
40	234	63	G1

For the group, b-y $0^m.28$-0.40 (F5-G1 stars), the range of variation of Δm_1 is 0.00-0.06. The metal content varies from that of the Hyades cluster to a value down by a factor of about 4 to 5.

The group b-y $0^m.22$-$0^m.27$ (F1-F4 stars) has a very small range of variations Δm_1. It is probable that the dispersion in metal content is very small in that group.

Near b-y = $0^m.22$, the dispersion in Δm_1 changes abruptly and the group b-y $0^m.05$-0^m21 (A3-F0 stars) shows a range Δm_1=0.06, with positive and negative values, indicating metal excess (metallic line stars).

C. Stellar Evolution.

1. Star Clusters and Stellar Evolution. A star cluster is a group of stars, physically related by their mutual gravitational attraction. The interest of such objects is that the members are likely to be about the same age. Detailed studies of a certain number of clusters with UBV photometry, has led to diagrams, similar to the H-R diagram, in which the stars of each cluster are distributed along narrow bands (Fig.4). The theory does not explain quantitatively all the details of the color-magnitude diagrams, but gives an explanation of the main features. Further development of the theory of the internal structure of the stars will

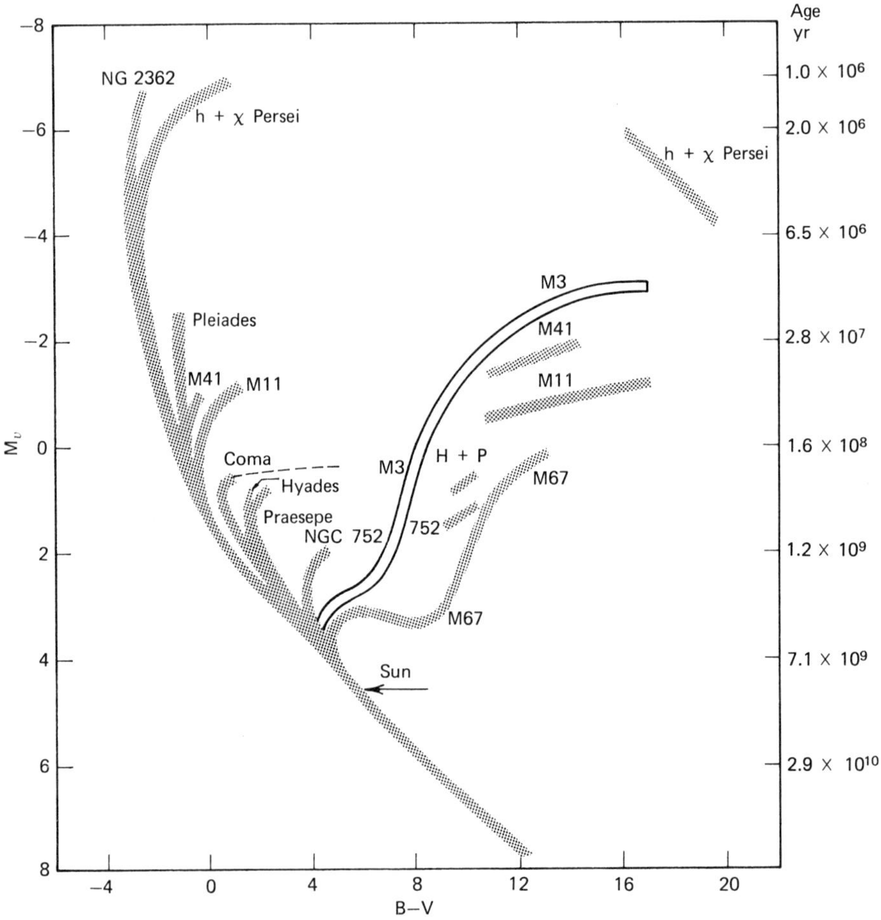

FIG. 4–The composite color-magnitude diagram for ten galactic and one globular cluster. The age of each cluster whose main-sequence termination point is at a given M_v is shown on the right. See text for details.

certainly improve the agreement between theory and observation.

Stars on the main sequence are of uniform chemical composition. Their evolution is so slow that they can be considered in a stationary state. The energy production is due only to thermonuclear reactions.

Preceding the main sequence phase, we have to consider the contraction towards the main sequence. During the contracting phase, the energy production is essentially due to the liberation of gravitational energy.

Following the main sequence phase, the stars become heterogeneous, as a consequence of the burning of the nuclear fuel in the central regions. The change of chemical composition is the reason for the evolution off the main sequence. During the evolution off the main sequence the main source of energy production is due to the thermonuclear reactions. However, gravitational contraction plays a non-negligible role during that phase. For a star of a given mass M, the stars begin to leave the main sequence when the nuclear fuel which has been burned is a fraction g(M) of the mass of the star. This gives the possibility of estimating the age of the star clusters by considering the mass of the stars which are just in the process of leaving the main sequence. We then obtain the following values for the age of different clusters (Table VII)

Table VII

Age of Galactic Clusters

Cluster	Age
NGC2362	1.0×10^{6} years
h and χ Per	2.0×10^{6}
Pleiades	28×10^{7}
M41	4×10^{7}
M 11	4×10^{7}
Coma	1.6×10^{8}
Hyades	4×10^{8}
NGC752	6×10^{8}
M67	7.1×10^{8}
NGC188	1.5×10^{10}

2. Star Associations. Star associations or aggregates are discovered when considering a selection of stars according to physical characteristics. The density of the associations is small, smaller than the local density of matter in the galaxy.

We can consider essentially the O associations, which consist of groups of stars of spectral type Bo or earlier. Their diameter varies from 30 to 200 parsecs. They are distributed in the galactic plane.

The associations are observed as a group of bright blue stars around one or several galactic clusters. For example, the association of Persei I is a system of 170 parsecs in diameter around two galactic clusters, the h and χ Per clusters, each of them having a diameter of 7 parsecs.

Ambarzumian has shown that the nuclei of the associations are always formed of an open cluster with O and BO stars. These open clusters always contain a small group of bright stars distributed in a Trapezium (like Orion Trapezium) or in a chain (like NGC7510, nucleus of Cas II).

Ambarzumian has shown that an important class of non-stable stars, the T Tau stars, always belong to an O association, the most remarkable example being the association of Orion. To the same family of non-stable stars, should be joined the flare stars. Both types will be described in more detail later. For the time being, it is sufficient to consider that these stars are irregular variables showing sudden increases in brightness. The increase has short duration in the case of the flare stars, which are much like K and M dwarfs, as far as their absorption spectrum is concerned. The increase is less frequent, and has a greater duration in the case of T Tau stars; their absorption spectrum ranges from F to M, the majority of them being F. T Tau stars are characterized by Hα and H and K of Ca II in emission and by several other spectroscopic features described by Herbig (1958).

An important remark of Ambarzumian (1954) was that T Tau stars are very young stars, their age being comparable to the age of the O stars of the association to which they belong. An estimate of the ages of such extremely young clusters can be given (Table VIII).

Table VIII

Ages of Extremely Young Clusters

Cluster	Age(10^6 years)
NGC2264	2.1
NGC6530	3
Orion	3
TC5146	3
NGC6611	1.8
h+χ Per	6

3. Stellar Evolution. Stellar evolution is usually understood to be the study of internal constitution of stars, and of the irreversible processes which take place inside them. However, (in the following) we shall be concerned essentially with the effect of phenomena on stellar evolution, that take place on the surface of the stars. Consideration of surface phenomena began with an attempt by Burbidge and Burbidge (1955) to explain the peculiar abundances in the atmosphere of peculiar A stars like α2CnV. In their first papers they considered the possibility of nuclear reactions (due to the local appearance of very hot spots) taking place in the outer layers of a star. At the same time, Parker (1955) suggested that a dynamo mechanism, due to hydrodynamic turbulence associated with differential rotation, could regenerate the magnetic field of the sun.

As early as 1950, Struve drew attention to the peculiar distribution of equatorial velocities along the main sequence. The possibility of explaining the loss of angular momentum by surface mass loss was studied by Massevitch (1949) and Fessenkoff (1949). Schatzman (1953-1954) tried to explain the distribution of angular velocities among stars of the main sequence by the mechanism of Massevitch and Fessenkoff. However, the mass to be lost from the stellar surface was found too large, of the order of 30% of the mass. Whereas the T Tauri stars seem to present a high rate of mass loss according to Kuhi (1964), stars on or near the main sequence do not show any spectral peculiarities which could be attributed to a high rate of mass loss.

In 1958 we faced a situation in which a variety of facts

could be attributed to surface phenomena, i.e., nuclear reactions, magnetic fields, statistical distribution of angular momentum, but a number of difficulties remained to be explained. Some of these difficulties are now understood, but in many cases a quantitative theory of the phenomena is still to be found. This is not surprising, since the evolutionary processes which we shall have to consider are based essentially on turbulence and accelerating mechanisms. Both are stochastic processes which certainly are not entirely understood at the present time.

The key to the whole question of surface phenomena is in the existence of the hydrogen convective zone of the stars. The convective zone is the source of turbulence and is the place where the waves that travel through the atmosphere are generated; it generates also the magnetic field of the stars from which stems the surface activity. We can see, therefore, that the understanding of the H CZ is also the key to the understanding of a number of aspects of stellar evolution.

In the following we shall consider and sketch the physical processes in which we are interested. We shall try to develop as far as possible the theory of all processes of which we have a fair understanding and draw attention to the difficulties which are not yet understood.

BIBLIOGRAPHY

Chapter I

AMBARZOUMIAN, V. A., (1954), Stellar Evolution: Discours introductif, Trans. I. A. U., VIII, 665.

BABCOCK, H. W., (1958), Magnetic Fields of the A-Type Stars, Ap. J., 128, 228.

BERTAUD, Ch., (1959), Catalogue d'étoiles, Ap. J. Obs., 42, 45.

BERTAUD, Ch., (1960), Catalogue d'étoiles, Ap. J. Obs., 43, 129.

BURBIDGE, G. R., and BURBIDGE, E. M., (1955), An Analysis of the Magnetic Variable α2 Canum Venaticorum, Ap. J. Suppl., 1, 431.

FESSENKOFF, V. G., (1949), On the Problem of Cosmogony in Contemporary Astronomy, A. J. U.S.S.R., 26, 67 (in Russian).

HERBIG, G., (1958), "Stars of Types F, G, and K, with Emission Lines; Introductory Report," in Stars with Emission Lines, 8th Liege Symposium, Mem. Soc. Roy. Sci. de Liege, Vol. XX, 251.

JASCHEK, C., and JASCHEK, M., (1958), The Position of the Peculiar A-Type Stars in the Color-Absolute Magnitude Diagram, Zs. f. Ap., 45, 35.

JASCHEK, C., and JASCHEK, M., (1962), Considerations on the Evolution of Peculiar and Metallic Line Stars, La Plata Symposium on Stellar Evolution, 137.

KEENAN, P. C., (1963), Classification of Stellar Spectra, in Basic Astronomical Data, (The University of Chicago Press), 78.

KRAFT, R. P., (1965), Studies in Stellar Rotation, I. Comparison of Rotational Velocities in the Hyades and Coma Clusters, Ap. J., 142, 681, Errata p. 1699.

KRAFT, R. P., (1965), Studies in Stellar Rotation, II. The Effect of Rotation on the Colors and Magnitudes of A- and F-Type Stars in the Hyades, Ap. J., 142, 703.

KUHI, L. V., (1964), Mass Loss from T Tauri-Stars, Ap. J., 140, 1409.

MASSEVITCH, A. G., (1949), Stellar Evolution Accompanied by Corpuscular Radiation from the Standpoint of the Theory of Internal Stellar Structure, A. J. U.S.S.R., 26, 207, (in Russian).

PARKER, E. N., (1955), Hydromagnetic Dynamo Models, Ap. J., 122, 293.

PECKER, J. C., and SCHATZMAN, E., (1959), Astrophysique Générale, (Paris: Masson et C.--Editeurs).

SARGENT, W. L. W., (1964), The Atmospheres of the Magnetic and Metallic Line Stars, Annual Rev. Ast. and Ap., 2, 297.

SCHATZMAN, E., (1954a), Remarques sur l'évolution des étoiles III, Interpretation des rotations axiales, Ann. d'Ap., 17, 300.

SCHATZMAN, E., (1954b), Remarques sur l'évolution des étoiles V, Sur la formation des systèmes d'étoiles, Ann. d'Ap., 17, 382.

SLETTEBAK, A., (1954), The Spectra and Rotational Velocities of the Bright Stars of Draper Types, B8-A2, Ap. J., 119, 146.

STROMGREN, B., (1963), Quantitative Classification Methods, in Basic Astronomical Data, (The University of Chicago Press), 123.

STRUVE, O., (1950), Stellar Evolution, Princeton University Press.

STRUVE, O., (1959), Elementary Astronomy, Oxford University Press.

WALLERSTEIN, G., (1962), Abundancies in G Dwarfs VI, A Survey of Field Stars, Ap. J. Suppl., 6, 407.

II. STRUCTURE OF THE OUTER LAYERS OF A STAR

A. The Convective Zone.

The classical result due to K. Schwarzschild (1905) is that an atmospheric layer with no viscosity becomes convectively unstable as soon as the following condition is fulfilled:

$$\left(\frac{d\log\rho}{d\log P}\right)_* < \left(\frac{d\log\rho}{d\log P}\right)_{ad} \tag{2.1}$$

When the chemical composition is uniform, this can be changed to:

$$\left(\frac{d\log T}{d\log P}\right)_* > \left(\frac{d\log T}{d\log P}\right)_{ad} \tag{2.2}$$

From an elementary study of the equations of motion in a plane parallel atmosphere, it can be readily seen that the quantity

$$\left(\frac{d\log\rho}{d\log P}\right)_* - \left(\frac{d\log\rho}{d\log P}\right)_{ad} = \Delta\nabla\rho \tag{2.3}$$

is related to the driving force of the motion. If $\Delta\nabla\rho$ is positive, it acts as a restoring force and the motion is stable; if $\Delta\nabla\rho$ is negative, the motion is dynamically unstable.

Let us consider the equations of hydrodynamics in the Lagrange form. If $\vec{r}$ is the Lagrangian coordinate, we have for the equation of continuity:

$$\rho\,\vec{\nabla}.\vec{r} = \rho_o \tag{2.4}$$

and for the equation of motion:

$$\rho\frac{d^2\vec{r}}{dt^2} = -\vec{\nabla}p - g\rho \tag{2.5}$$

Let us assume that the displacement is adiabatic, in other words that:

$$\frac{dp}{dt} = \frac{\gamma p}{\rho}\frac{d\rho}{dt} \quad . \tag{2.6}$$

To the approximation of the small motion, we have

$$\vec{r} = \vec{\sigma}_o + \vec{\xi} \tag{2.7}$$

and $\vec{\xi}$ is small. We have then:

$$\rho = \rho_o - \rho_o \vec{\nabla}_o \cdot \vec{\xi} \tag{2.8}$$

$$\sigma \frac{d^2\vec{\xi}}{dt^2} = -\nabla_o p_o + (\nabla\xi\nabla) p_o + \vec{\nabla}_o \gamma p_o \vec{\nabla}_o \cdot \vec{\xi} - \vec{g}\rho_o + \vec{g}\rho_o \vec{\nabla}_o \cdot \vec{\xi} \tag{2.9}$$

$$p = p_o - \gamma p_o \vec{\nabla}_o \cdot \vec{\xi} \tag{2.10}$$

Elimination of ξ gives an equation for dilatation S :

$$\left\{\left(\frac{\partial^2}{\partial t^2} + g \frac{\partial}{\partial z}\right)\left(\frac{\partial^2}{\partial t^2} - c^2\nabla^2 + (\gamma-1)\, g \frac{\partial}{\partial z}\right) + g\nabla^2\left(c\frac{2\partial^2}{\partial z^2} - (\gamma-1)g\right)\right\} s = o \tag{2.11}$$

If we take a short horizontal wave length and a high frequency, we can apply the WKB approximation to find

$$\omega^2 = \frac{g^2(\gamma-1)}{c^2}\left(1 + \frac{1}{(\gamma-1)g}\frac{\partial c^2}{\partial z}\right) \tag{2.12}$$

The quantity $\frac{1}{(\gamma-1)g}$ is just $\left(\frac{-\partial c^2}{\partial z}\right)_{ad}$ and we can write

$$\omega^2 = \frac{g^2(\gamma-1)}{c^2}\left\{1 - \frac{\left(\frac{\partial c^2}{\partial z}\right)*}{\left(\frac{\partial c^2}{\partial z}\right)_{ad}}\right\} \tag{2.13}$$

We see that when condition (2) is fulfilled, ω^2 is negative and there is convective instability.

An important quantity is the Reynolds Number. The Reynolds Number is equal to the ratio of the macroscopic viscosity ℓv to the microscopic viscosity λv_{th}, where ℓ is a characteristic length, v the velocity of the macroscopic motion, λ the mean free path, v_{th} the thermal velocity. We have

$$R = \frac{\ell v}{\lambda v_{th}} \tag{2.14}$$

In the convective zone, (v/v_{th}) can be small, of the order of 10^{-1} or 10^{-2} ; it is unlikely to be very much smaller. On the contrary, λ in the stellar plasma is very small, of the order of 10^{-2} cm to 10^{-6} cm. Any reasonable value of ℓ is of the order of 10 km at least. The Reynolds number is therefore at least 10^{6}, and can be as large as 10^{12}. For such large values, the motion is certainly turbulent. But, though the case of the thermally driven turbulence is quite clear cut, the theory of turbulence is so far from the known phenomena that it is not surprising that the theory of the stellar convective zone is still very primitive.

The properties of homogeneous turbulence cannot even be applied, as the size of the convective zone is always large compared to H,

$$H=\left(\frac{dz}{d\log p}\right) \quad . \tag{2.15}$$

H is called the scale height. When the thickness of the convective zone is small compared to H, the Boussinesque approximation can be applied. In that case, the compressibility can be neglected, except in the expression of the bouyancy force, in which case the turbulence can be considered as taking place in an incompressible fluid. This is unfortunately not the case in stars, and we have always (or almost always)

$$(\Delta z)_{conv} \gg H \tag{2.16}$$

The theory of the convective zone is yet in the state of a mixing length theory.

Before briefly considering the properties of the convective zone as it results from the mixing length theory, we must make a statement concerning the statistical assumptions underlying the theory.

When we consider a spherically symmetric star, we average the properties of the convective zone over a sphere, in order to be able to introduce average physical quantitites which depend only on the radius of the sphere: this includes pressure, density, temperature, entropy.

As a first example, let us consider the expression of the flux. To begin with we shall make a simplifying assumption.

If we can assume that in all of the convective zone, including its upper boundary, the optical thickness is great, we then can write the equation of conservation of flux

$$\sigma T_{eff}^4 = \frac{16}{3} \frac{\sigma T^4}{\bar{k}\rho H} + F_{conv} \tag{2.17}$$

where $\bar{k}$ is the Rosseland mean of the absorption coefficient, σ is the constant of Stefan and $H = (dh/d \log P)$ is the local equivalent height. The first term represents the radiative flux, and is also a type of average, taken over the sphere.

As soon as the local radiative flux drops below the total flux, the remaining part of the flux is transported by convection. This effect is due essentially to the rapid growth of the absorption coefficient with temperature in the outer layers of the star.

Deeper inside the convective zones, however, the absorption coefficient decreases and convection stops at the point where the local radiative flux again becomes equal to the total flux. This is only an alternative way of expressing the Schwarzschild condition for the boundary of the convective zone.

A critical quantity is the depth of the convective zone.

To disentangle the physics of the problem, we can consider a radiative atmosphere with constancy of flux and an adiabatic convective zone fitted to the atmosphere. We shall, for the time being, limit ourselves to main sequence stars. Following the adiabatics, we stop the convective zone at the place where the Schwarzschild condition is fulfilled. Even with such a crude approximation, it is easy to see that an A star has a very shallow convective zone and a G star has a deep convective zone; this difference in structure is attributed to the behavior of the absorption coefficient.

We cannot be satisfied with such a crude model, since a

very important question is the exact determination of the convective zone's depth as a function of the spectral type. Eventually, we also want to build a grid of models for various gravity and effective temperatures, and to carry the representation of the grid of models in the Hertzsprung-Russell diagram.

Let us first consider the standard approximation which is used for studying the convective zone:

The convective flux is expressed with the help of the mixing length theory. This implies the following approximations:

(a) Any change over a distance ℓ is expressed in a linearized theory; for example, for the temperature,

$$\delta_i T = \left(\frac{d \log T}{d \log P}\right)_i \cdot \frac{\ell}{H} \tag{2.18}$$

where the index i refers to a specified kind of change of T.

(b) The enthalpy of the fluid is supposed to be carried by bubbles of matter which exchange enthalpy at constant pressure after having moved over a distance ℓ. This explains the introduction of the heat capacity at constant pressure C_p;

(c) The change of enthalpy during motion over a distance ℓ at velocity v includes the exchange of energy by radiation during motion.

(d) The mixing length ℓ is usually taken to be equal to once or twice the local equivalent scale height.

Let us now go through the calculations. If T is the temperature in the bubble, T* the average temperature, the heat flux per unit surface transported by the ascending currents is

$$w \rho C_p \ell \left(\frac{dT}{dz} - \frac{dT^*}{dz}\right) \tag{2.19}$$

where w is the average velocity of the currents, and ℓ the mixing length. Assuming that the ascending currents occupy one-half of the surface, we get

$$\Phi_c = 1/2\, w \rho C_p \ell \left(\frac{dT}{dz} - \frac{dT^*}{dz}\right) \tag{2.20}$$

The average velocity w of the ascending currents can be obtained by using the linearized theory of the ascending motion under the bouyancy force:

$$\rho \frac{d^2 z}{dt^2} = \left(\rho^* - \rho\right) g \tag{2.21}$$

or

$$\rho \frac{d^2 z}{dt^2} = g\, z \left(\frac{d\rho^*}{dz} - \frac{d\rho}{dz}\right) \tag{2.22}$$

using the relation

$$\frac{1}{\rho}\frac{d\rho}{dz} = \frac{1}{P}\frac{dP}{dz} + \frac{1}{\mu}\frac{d\mu}{dz} - \frac{1}{T}\frac{dT}{dz} \tag{2.23}$$

we obtain the linear approximation

$$\frac{d\rho^*}{dz} - \frac{d\rho}{dz} = \frac{\rho}{T}\left(\frac{dT}{dz} - \frac{dT^*}{dz}\right) Q \tag{2.24}$$

with

$$Q = \left(1 - \frac{T}{\mu}\frac{d\mu}{dT}\right) \tag{2.25}$$

We derive the velocity $\frac{dz}{dt} = z'$,

$$z' = z \left(\frac{g}{H}(\nabla^* - \nabla)Q\right)^{1/2} \tag{2.26}$$

The average value w is related to z' at the distance ℓ by the relation

$$w^2 = a\, \frac{g\ell^2}{H}(\nabla^* - \nabla)Q \tag{2.27}$$

where a is a numerical quantity.

If we take the average

$$w = \frac{1}{\ell} \int_0^{\ell} z' \, dz \tag{2.28}$$

we obtain a = 1/4.

We have now to determine the actual gradient ∇ in the bubble motion, compared to the adiabatic motion. We can express in two different ways the ratio of the energy exchanged at distance ℓ to the energy radiated away during motion.

If the motion were adiabatic, the gradient would be ∇_{ad}. It is, instead, ∇. The difference $\nabla - \nabla_{ad}$ expresses the amount of energy lost by radiation. The energy exchanged, as we have seen, is proportional to $(\nabla^* - \nabla)$.

Due to a difference of temperature δT, the energy exchanged is $C_p \, \rho \, \delta T$ x volume; similarly we derive the expression of the energy radiated away during time (ℓ/w). From the expression of the flux

$$F = \frac{4\pi}{3} \frac{dB}{dt} \tag{2.29}$$

we derive the expression

$$F = \frac{4}{3} \sigma T^4 \frac{4}{\bar{\mathcal{H}} \rho \ell} \frac{\delta T}{T} \tag{2.30}$$

and finally,

$$\frac{\nabla^* - \nabla}{\nabla - \nabla_{ad}} = \frac{C_p \, \rho \, \delta T \text{ x volume}}{\frac{16}{3} \sigma T^4 \frac{1}{\bar{\mathcal{H}} \rho \ell} \frac{\delta T}{T} \frac{\ell}{w} \text{ surface.}} \tag{2.31}$$

For a spherical bubble, this gives

$$\gamma = \frac{\nabla^* - \nabla}{\nabla - \nabla_{ad}} = \frac{1}{16} \frac{\rho^2 C_p \mathcal{H} \ell \, w}{\sigma T^3} \tag{2.32}$$

The Peclet Number expresses the ratio of the heat transported by convection to the heat transported by radiation:

$$\underline{P}_e = \frac{w\ell}{(\lambda / C_p)} \tag{2.33}$$

where

$$\lambda = \frac{16\pi\sigma\, T^3}{3\,\bar{\mathcal{H}}\rho}$$

is the radiative heat conductivity.

We have

$$\gamma = \frac{P_e}{3} \tag{2.34}$$

Technically, the actual gradient ∇^* is given by a cubic equation. Introducing ∇_R, defined to be the gradient which would exist if all the energy were transported by radiation,

$$\Phi = \frac{16}{3}\,\frac{\sigma\, T^4}{\mathcal{H}\rho H}\,\nabla_R \tag{2.35}$$

and the dimensionless quantity D

$$D = \frac{\sigma\, T^3\, H}{C_p \rho^2 \mathcal{H} \ell^2}\left(\frac{\mu}{a\ RTQ}\right)^{1/2} \tag{2.36}$$

we obtain the two equations:

$$(\nabla^* - \nabla)^{1/2} = \frac{1}{2\ D}\,(\nabla - \nabla_{ad}) \tag{2.37}$$

$$(\nabla^* - \nabla_R) + \frac{3}{4\ D}\,(\nabla^* - \nabla)^{3/2} = 0 \tag{2.38}$$

from which we derive the equation:

$$(\nabla^* - \nabla_{ad}) - (\nabla_R - \nabla_{ad}) + \frac{3}{4D}\left\{\left((\nabla^* - \nabla_{ad}) + D^2\right)^{1/2} - D\right\}^3 = 0 \tag{2.39}$$

which is a cubic equation in $(\nabla^* - \nabla_{ad} + D^2)^{1/2}$.

Baker and Temesvary (1966) compute a radiative atmosphere to optical depth 2/3, and then begin the calculation of the solution.

What is interesting in the results is the following:

(a) The depth of the convective zone depends on the mass and luminosity of the star.

(b) The depth of the convective zone depends on the assumption concerning the value of the mixing length ℓ. Tables are given by Baker and Temesvary for two values of $\alpha = \ell/H$, $\alpha = 1$, and $\alpha = 2$. The greater the mixing length, the deeper the convective zone.

As we shall refer extensively to the properties of the convective zone, we have first to consider the gross properties of the HCZ. To that effect, we shall draw a map giving the extension in depth of the convective zone in the log T_{eff}, log g plane (Fig. 5).

Similarly, it is possible to draw a line in the H-R diagram (Fig.6) showing the separation between the region of a deep convective zone and the region with a shallow convective zone, or none.

The problem of the place of the lower boundary of the convective zone is important for stellar evolution, as thermonuclear reactions of light elements are very sensitive to the temperature. A small increase in temperature can make the burning of lithium possible during a stellar lifetime.

Still in the frame of the mixing length theory, several improvements can be made. Observations of turbulence in water, due to thermal instability (Fig.7a), suggest that ℓ can be taken equal to the distance z to the nearest boundary in all the HCZ. Another possiblity is to take ℓ equal to the distance z to the nearest boundary as long as $z \leq H$, and then to take $\ell = H$.

Böhm (1966) and Kohl (1966) have extended the theory of the convective zone to include that approximation, bringing the convective zone closer to a constant specific entropy, and thus placing the lower boundary at a deeper level.

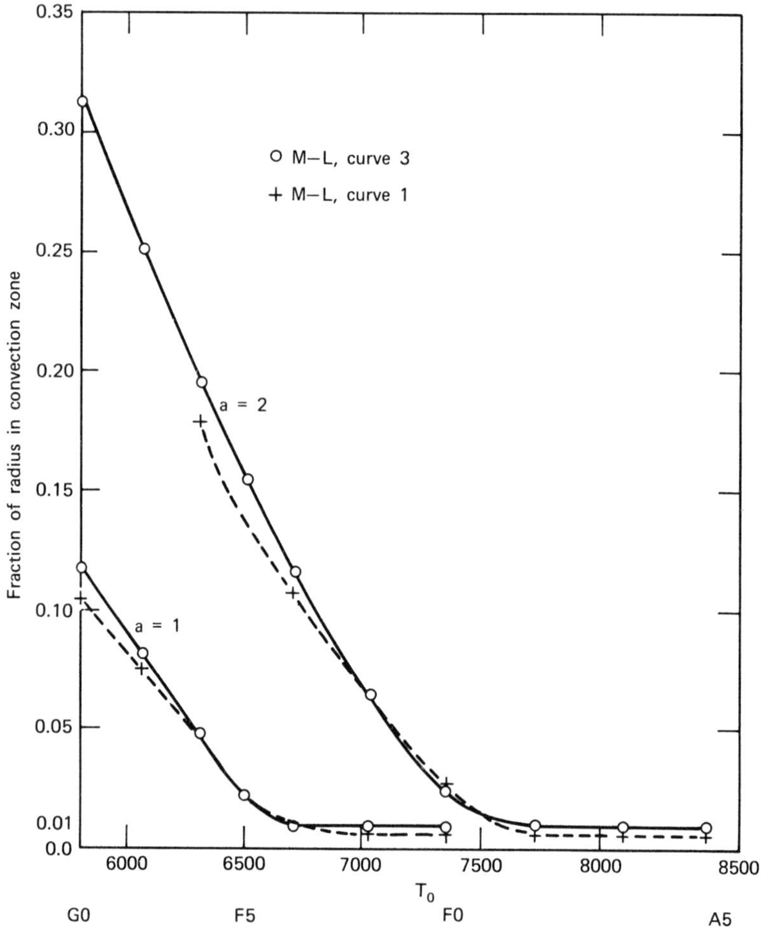

FIG. 5–Depth of outer convection zone in main sequence stars. (Courtesy N. Baker and R. Kippenhahn).

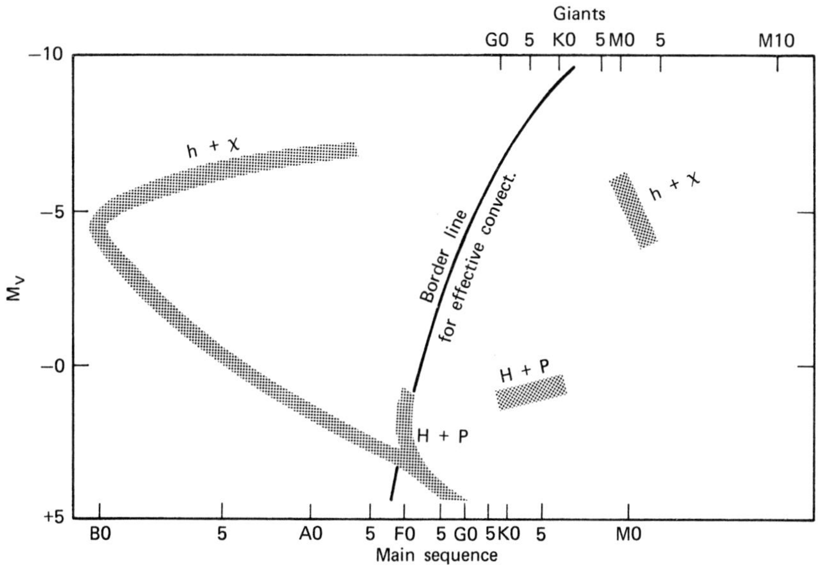

Fig. 6 Colour-magnitude diagram for population I stars (Hyades, Pleiades, $h + \chi$ Persei) with an approximate dividing line for effective convection (——). Above this line no effective convection is expected theoretically.

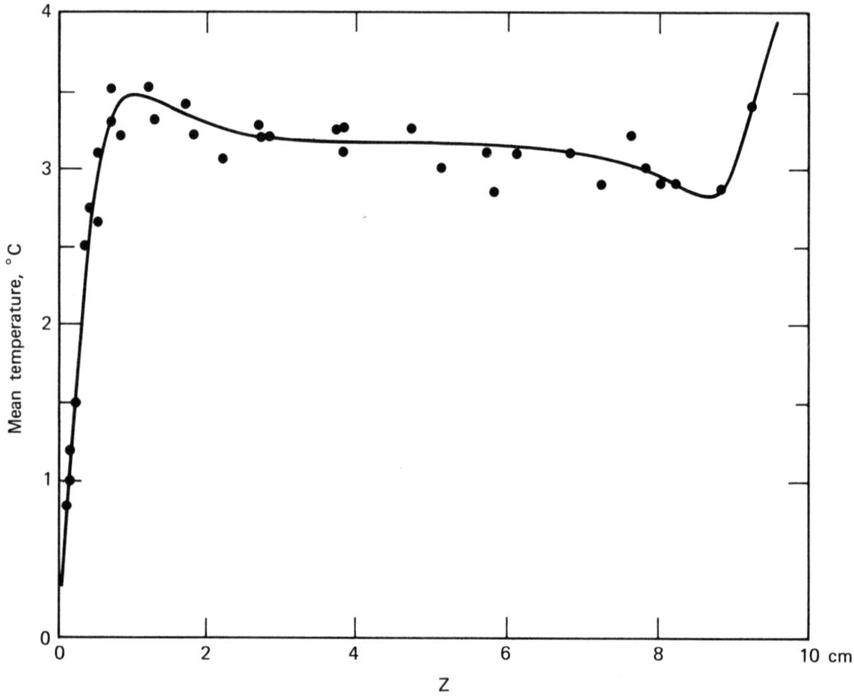

Figure 7a

It is important to consider more closely the sum of physical quantities in a convective zone, built in the mixing length theory. Let us consider first the density mass, the problem of the velocity field being delayed to the next section.

It can be inferred from the known results in the ($\log g$, $\log T_{eff}$) plane, there is a region where a density inversion takes place in the outer layers. If it were true, this would lead to the occurrence of a Rayleigh-Taylor instability.

Let us consider, for example, one of the models calculated by Mihalas (1965) for which we give the distribution of the density as a function of depth. (Fig. 7 b and c). We see that density has a minimum about 200 km below the maximum.

It is quite probable that the density inversion cannot exist as a permanent feature of the atmospheric structure, whereas the average density can very well be replaced by a constant until the point is reached where the density can begin to rise. It is not at all unlikely, considering the size of the convective zone, that temporary density inversions are established followed by a mixing at constant density. As we shall see later, this can lead to an elementary explanation of the microturbulence in the A_m stars.

The exact fitting to the radiative envelope has been considered in great detail by Ch. Pecker-Wimel (1953) in the case of an adiabatic convective zone. More recently, Sack (1966) considered the exact fitting of the radiative envelope to the convective zone. Since the convective zone becomes almost adiabatic at great depths, Sack considers the specific entropy exclusively as a function of gravity and effective temperature.

The expression <u>exact fitting</u> is not satisfactory in the frame of the mixing length theory. It is impossible to make an exact fit, since the radiative transfer is expressed exactly in the radiative zone, but is simplified in the convective zone to correspond to the case of great optical depths. The expression for radiative flux in the radiative zone is:

$$F_R = 2\pi\left[\int_\tau^\infty S(t)\, K_2(t-\tau)\, dt - \int_0^\tau S(t)\, K_2(\tau-t)\, dt\right] \tag{2.40}$$

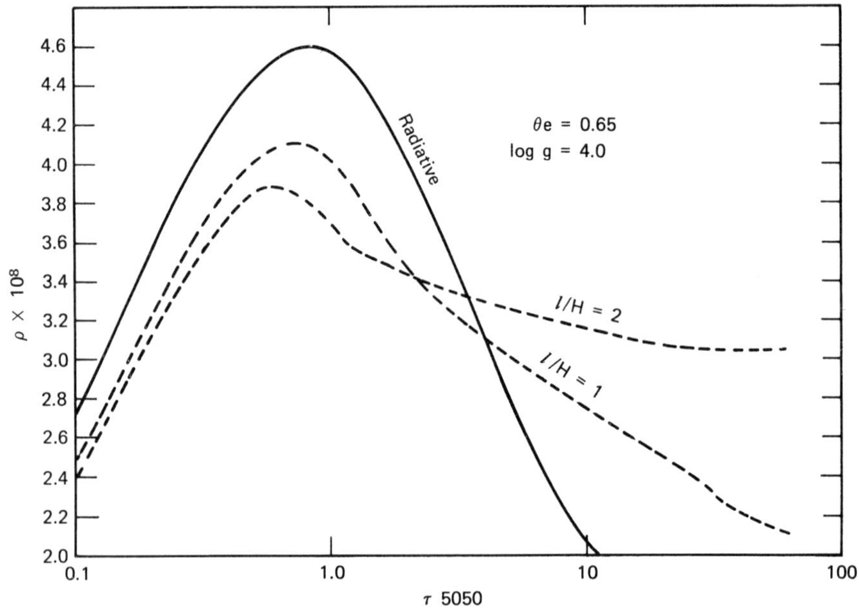

FIG. 7b–Run of density with depth showing density inversions characteristic of the mixing theory.

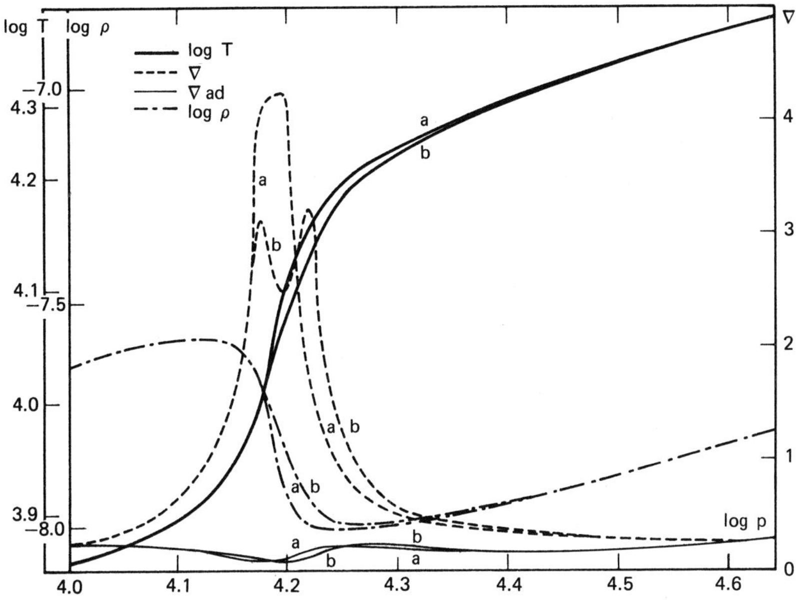

FIG. 7c–Comparaison du modèle A M2 (*a*) et du modèle A M2 (*b*).

and in the convective zone it is taken as:

$$F_R = \frac{16}{3} \quad \frac{\sigma T^4}{k\rho H} \tag{2.41}$$

An exact fit is naturally impossible.

Further improvements can be considered in the frame of the mixing length theory:

(a) It would naturally be possible to abandon the linearized theory of the heat transport and to use an integral form for the amount of heat transported from level x to the boundary of the convective zone:

$$F_{conv} = \rho C_p v \int_x^z (\nabla^* - \nabla) \, T \, \frac{dx}{H} , \tag{2.42}$$

in which ∇ is the average gradient and ∇^* the gradient associated with the bubble. This is probably as far as we can go with the mixing length theory, at least for solutions of heat transport.

The same integral form could be used for the conservation of energy:

$$\frac{1}{2} \rho v^2 = \int_k^z g\rho d \, \xi \int_k^{\xi} (\nabla^* - \nabla) \, \frac{1}{H} \, (1 - \frac{\partial \log\mu}{\partial \log T}) \, d\eta \tag{2.43}$$

Locally, the bubble gradient and the actual gradient will still be related to the adiabatic gradient by the relation

$$\frac{\nabla^* - \nabla}{\nabla - \nabla_{ad}} = \frac{Pec}{3} , \tag{2.44}$$

with the Peclet Number defined locally as

$$Pec = \frac{F_{conv}}{F_{rad}} \tag{2.45}$$

The nature of the improvement is questionable. An attempt to solve the problem numerically is under way.

(b) A second improvement would be to write the radiative flux properly in the convective zone and use the solutions of the non-gray radiative transfer problem. But it seems that such an improvement should be delayed until a better approximation of the convective transport of energy is obtained. However, this is essentially the method followed by Mihalas (1965).

B. The Velocities and Frequencies.

In the frame of the mixing length theory, with a mixing length equal to α scale height, the characteristic velocities are given by:

$$w^3 = 2 \frac{a\ RTQ}{\mu} \frac{(\ell/H)}{\rho TC_p} \Phi_c \qquad (2.46)$$

It is easy to calculate velocities and characteristic times $t \simeq (H/v)$. In the case of the sun, at a depth of one thousand kilometers, the velocity which is found is of the order of three kilometers per second and is near its maximum.

Some doubt should be raised about the validity of this expression for the characteristic velocity, as it leads to velocities smaller than but comparable to the sound velocity. There is no reason to believe that the theory of energy transport by bubbles applies at quasi-sonic velocities. The characteristic time is of the order of 100 seconds. Near the top of the convective zone, the velocity has dropped to about 1 kilometer per second, and the characteristic time has increased to three minutes. At greater depths the velocity decreases and the characteristic time increases; Bohm's estimate of the mixing time for the whole convective zone is of the order of five years.

These results have to be compared with the observations. In the case of the sun, Doppler shift and Doppler broadening can be observed; in the case of the stars, only Doppler broadening (microturbulence and macroturbulence) can be observed (micro and macro refer to the size of the turbulent elements compared to the mean free path of the photon). However, the interpretation of the observations is not obvious, as the spectral lines are formed in an extended region, including the convective zone and the radiative atmosphere above it.

This is the reason why an analysis of the motions generated in a stellar atmosphere has to be conducted with some care. We shall consider first the motions generated by the pressure variations and later the case of the atmospheres with an inversion of density.

The velocity fluctuations behave like a source of pressure waves. We, therefore, have waves propagating in the stable zone of the stellar atmospheres, and the waves reflect the properties of the turbulence which has given birth to them.

The spectral lines show certainly the effect of both the turbulent motion in the convective zone and the chaotic field of pressure waves in the radiative zone. However, as the spectral lines are generally formed fairly high in the atmosphere (especially the moderately strong lines which are essential for the determination of the microturbulence), the properties of the motions due to the pressure waves are essential for the understanding of the line formation.

First of all, let us consider elementary properties of wave propagation.

We shall be concerned more with the way in which waves produced by the pressure fluctuations reach the outer layers. To that end, we shall consider first the properties of the diagnostic diagram. The following is inspired by the recent work of P. Souffrin (1966).

If we consider the linearized equations of propagation, we consider solutions of the form,

$$v = v(z) \exp i\,[Kx + \ell y + \omega t - S(z)]. \qquad (2.47)$$

The equation of dispersion, for a given value of k, ℓ, and ω determines S and tells us whether we are concerned with a travelling wave or a standing wave. Due to the condition that the amplitude of the wave decreases sufficiently fast at infinity, the only kind of standing waves we need to consider are evanescent waves.

Such an effect exists even in the case of a uniform fluid. Let us consider the operator

$$L = \frac{1}{c^2} \frac{\partial^2}{\partial t^2} - \nabla^2, \tag{2.48}$$

and a solution of the form

$$\exp i\,(kx + \omega t - mz),$$

this gives

$$\frac{\omega^2}{c^2} = k^2 + m^2, \tag{2.49}$$

or

$$m^2 = \frac{\omega^2}{c^2} - k^2 . \tag{2.50}$$

We can see immediately that a source with

$$k^2 > \frac{\omega^2}{c^2} \tag{2.51}$$

does not produce propagating waves. The curve $m = o$, or $c\,k = \omega$ divides the diagnostic diagram into two regions (Fig. 8), one in which waves propagate, and one in which they do not propagate.

The amplitude of a wave, $A(\underline{k},\omega)$, experiences a linear transform $A(k,\omega)f(k,\omega,z)$ through propagation, which is easily found to be:

$$|f|^2 = 1, \text{ if } k < \frac{\omega}{c},$$

$$|f|^2 = \exp\left[-2z\sqrt{k^2 - \frac{\omega^2}{c^2}}\right] \text{ if } k > \frac{\omega}{c} . \tag{2.52}$$

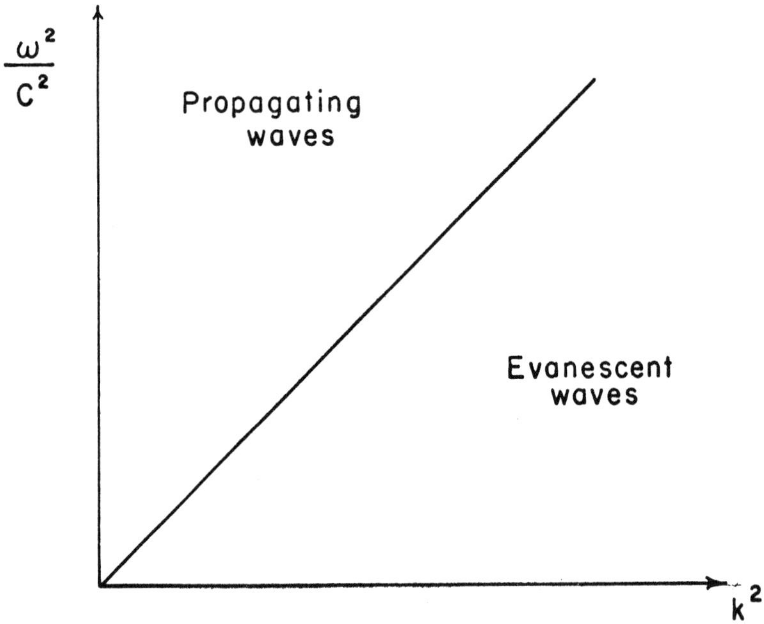

Figure 8

The situation is also well known for isothermal atmospheres where the diagnostic diagram shows two propagating regions; one for the pressure waves, the other for the gravity waves (Fig. 9). Two characteristic frequencies appear; one is due to the reflection of pressure waves by the density gradient, and is given by

$$\omega_c^2 = \frac{c^2}{4H^2} . \tag{2.53}$$

The other characteristic frequency corresponds to undamped gravity waves, and is known as the Väisälä frequency,

$$\omega_v^2 = \frac{g}{H} - \frac{g^2}{c^2} = 4 \frac{\gamma - 1}{\gamma^2} \omega_c^2 . \tag{2.54}$$

We shall consider separately the problem of the high frequency pressure waves (propagating waves) and the problem of low frequency waves (essentially, propagating gravity waves).

For the high-frequency part of the spectrum, which can be considered as a noise (in the acoustic sense of the word), we are interested in the power which leaves the convective zone.

We shall follow here essentially the work of Unno (1964), which derives its essential ideas from the work of Lighthill (1952). To this effect, let us consider the equation of motion, the equation of continuity, and the equation of energy in a uniform medium:

$$\frac{\partial}{\partial t} (\rho u_i) + \frac{\partial}{\partial x_j} (\underline{p} \, \delta_{ij} - \tau_{ij} + \rho u_i u_j) = \rho F_i \tag{2.55}$$

$$\frac{\partial \rho}{\partial t} + \frac{\partial}{\partial x_j} (\rho u_j) = q \tag{2.56}$$

$$\frac{\partial p}{\partial t} + u_i \frac{\partial p}{\partial x_i} = c^2 \left(\frac{\partial \rho}{\partial t} + u_i \frac{\partial \rho}{\partial x_i} \right) \tag{2.57}$$

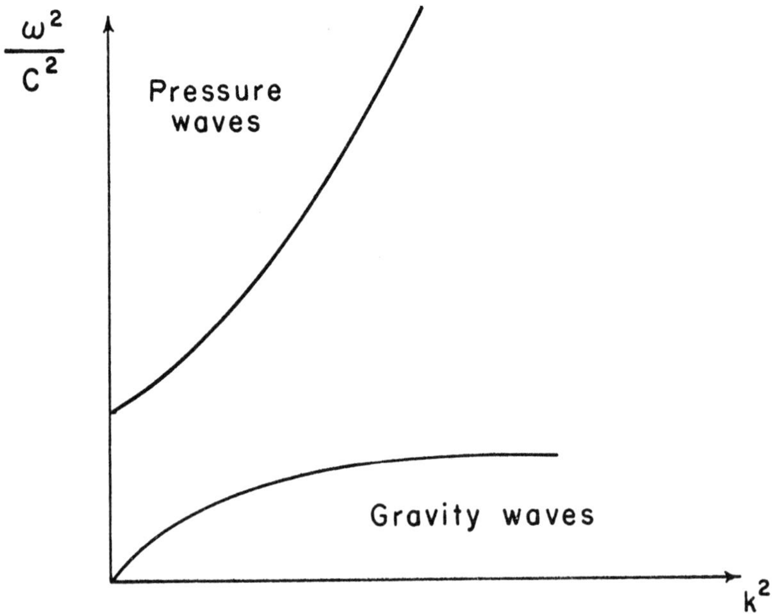

Figure 9

where u_i is the velocity; $\underline{P}$, the pressure
ρ, the density
τ_{ij}, the viscous tensor
F_i, the external forces applied to the system
q, a source term
$c^2 = \frac{\gamma p}{\rho}$ is the sound velocity

An interesting idea of Lighthill's is to make a proper choice of variables such as to separate the source terms from the propagating terms. As we know the form of the propagating term, and that the propagation can be entirely ascribed to the density fluctuations, we can combine equations(2.55)and(2.56)after differentiation by $\frac{\partial}{\partial x_i}$ and by $\frac{\partial}{\partial t}$, to obtain

$$\frac{\partial^2 \rho}{\partial t^2} = \frac{\partial q}{\partial t} - \frac{\partial}{\partial x_j}(\rho F_j) + \frac{\partial}{\partial x_i \partial x_j}(P \delta_{ij} - \tau_{ij} + \rho u_i u_j) \tag{2.58}$$

In order to introduce the propagation term, subtract $\frac{\partial}{\partial x_j} c^2 \frac{\partial}{\partial x_j} \rho$ from both sides:

$$\frac{\partial^2 \rho}{\partial t^2} - \frac{\partial}{\partial x_j} c^2 \frac{\partial \rho}{\partial x_j} = \frac{\partial q}{\partial t} - \frac{\partial}{\partial x_j}\left[\rho F_j + \frac{\partial c^2}{\partial x_j}\right] + \frac{\partial^2}{\partial x_i \partial x_j}\left[(p - \rho c^2)\,\delta_{ij} - \tau_{ij} + \rho u_i u_j\right] \tag{2.59}$$

The right hand side of the equations can be considered as a source term.

The $\frac{\partial q}{\partial t}$ term represents a monopole source, the $\frac{\partial}{\partial x_j}\left[\rho\left(F_i + \frac{\partial c^2}{\partial x_j}\right)\right]$ represents a dipole source,

the $\frac{\partial^2}{\partial x_i \partial x_j}\left[(p - \rho c^2)\,\delta_{ij} - \tau_{ij} + \rho\, u_i u_j\right]$ represents a quadrupole source.

The solution of the equation of propagation with a right hand side:

$$\frac{\partial^2 f}{\partial t^2} - c^2\nabla^2 f = -4\pi g\,(x,t) \tag{2.60}$$

has a classical solution with the retarded potentials:

$$G = \frac{1}{4\pi}\int d\vec{r}'\,\frac{g\left(\vec{r}', t - \frac{|\vec{r} - \vec{r}'|}{c}\right)}{|\vec{r} - \vec{r}'|} \tag{2.61}$$

The perturbation of the density is then (for the quadrupole term):

$$\rho - \rho_o = \frac{1}{4\pi c^2}\,\frac{\partial^2}{\partial x_i \partial x_j}\int T_{ij}\left(\vec{r}', t - \frac{|\vec{r} - \vec{r}'|}{c}\right)\frac{d\vec{r}'}{|\vec{r} - \vec{r}'|} \tag{2.62}$$

where we have written:

$$T_{ij} = (p - \rho c^2)\,\delta_{ij} - \tau_{ij} + \rho u_i u_j \tag{2.63}$$

The main term is $\rho u_i u_j$ due to the turbulence; the pressure term $(p - \rho c^2)$ would be exactly zero if the pressure variations were adiabatic.

At great distances from the source, the density variation is:

$$\rho - \rho_o = \frac{1}{4\pi c^2}\,\frac{r_i r_j}{x^3}\int_D \frac{1}{c^2}\,\frac{\partial^2}{\partial t^2}\,T_{ij}\left(\vec{r}', t - \frac{|\vec{r} - \vec{r}'|}{c}\right)dr' \tag{2.64}$$

The integration is made in the volume D which is the source of noise.

An order of magnitude of the noise produced can be obtained by a dimensional analysis. The derivative $\frac{\partial}{\partial t}$ will introduce a factor of the order of (u/ℓ), where u and ℓ are a characteristic velocity and a characteristic length of the turbulence. The tensor T_{ij} will be of the order of ρu^2, and the integration on the volume will give a factor of the order of ℓ^3. We then have

$$\rho - \rho_o = \frac{1}{4\pi c^2} \frac{1}{x} \frac{1}{c^2} \left(\frac{u}{\ell}\right)^2 \rho_o u^2 \ell^3 = \rho_o \left(\frac{u}{c}\right)^4 \frac{\ell}{x} \qquad (2.65)$$

The intensity of the sound is equal to the average value <pu>. The pressure variation is $c^2\delta\rho$, and the velocity variation is $c\,\frac{\delta\rho}{\rho}$, from which we derive the intensity $\rho_o u^8 c^{-5} \frac{\ell}{x}^2$. After integration on the sphere, the acoustic power radiated away is found to be

$$W \sim \rho_o c^3 \ell^2 \left(\frac{u}{c}\right)^8$$

the numerical coefficient depending on the properties of the turbulence. Lighthill (1961) has given experimental proofs of the result. In another form, as the power dissipated by the turbulence is of the order $\rho_o u^2 \frac{u\ell^3}{e} = \epsilon_T$, the acoustic power is

$$W \sim M^5 \epsilon_T$$

where M is the Mach number of the turbulence.

Unno (1964) has obtained the solution for the acoustic power in the case of an atmosphere. However, he does not obtain the solution for the general case, but only the solution for the isothermal atmosphere and isothermal disturbances. This has the inconvenience of suppressing the gravity waves, but the advantage of the propagation at constant velocity, and does not alter the main hydrodynamical properties.

By considering the variable

$$\varphi = p_o^{-1/2} \rho_1 (\vec{x}, t) \tag{2.66}$$

where ρ_1 is the perturbation of the density, Unno obtains

$$\left(\frac{\partial^2}{\partial t^2} - c^2 \Delta^2 + \frac{g^2}{4c^2} \right) \varphi (\vec{x}, t) = S(\vec{x}, t) \tag{2.67}$$

where

$$S = -p_o^{-1/2} \left[\vec{\nabla}_o \vec{f} + \left(\frac{\partial}{\partial t} \right)^{-1} \vec{g} \cdot \vec{\nabla}_q - \frac{1}{c^2} \left(\frac{\partial}{\partial t} \right)^{-1} \left(\frac{\partial^2}{\partial t^2} + \vec{g} \cdot \vec{\nabla} \right) h \right] \tag{2.68}$$

where $\vec{f}$, q, h have the following meaning:

$$\left. \begin{aligned} \vec{f} &= - \frac{\partial (\rho \vec{u}\, u_j)}{\partial x_j} - \frac{\partial (\rho_1 \vec{u})}{\partial t} \\ q &= - \vec{\nabla} \cdot (\rho_1 \vec{u}) \\ h &= - \gamma p_1 \vec{\nabla} \cdot \vec{u} - \vec{u} \cdot \vec{\nabla} p_1 \end{aligned} \right\} \tag{2.69}$$

where p_1 is the perturbation in pressure.

Neglecting terms of order higher than the second, and a propagating term, similar to the left hand side, Unno gives

$$S = p_o^{-1/2} \frac{\partial^2 (\rho_o u_i u_j)}{\partial x_i \partial x_j} = \frac{1}{c} \left[\frac{\partial^2 (\rho_o^{-1/2} u_i u_j)}{\partial x_i \partial x_j} - \frac{g}{c^2} \frac{\partial (\rho_o^{1/2} w u_j)}{\partial x_j} + \frac{g^2}{4c^2} (\rho_o^{1/2} w^2) \right] \tag{2.70}$$

where $\vec{g} = (o, o, -g)$ and $\vec{u} = (u_1, u_2, w)$.

The source term consists of three terms which differ from each other in the degree of space differentiation. They express respectively quadrupole, dipole, and monopole sources. It is essential to notice that the inhomogeneity gives rise to dipole and monopole sources.

The left hand side is the propagation term modified by the presence of the gravity. This introduces, naturally, by Fourier transform, the frequency σ_1:

$$\left.\begin{aligned} \sigma_1^2 &= \sigma^2 - \sigma_o^2 \\ \sigma_o^2 &= \frac{g^2}{4c^2} \end{aligned}\right\} \qquad (2.71)$$

with the radiation condition, $\sigma_1 \sigma > o$.

The solution is obtained for the Fourier amplitude $\rho_\sigma(\vec{r})$,

$$\rho_\sigma(\vec{r}) = \frac{1}{8\pi^2 c^2} \int_V \frac{d\vec{r}\,'}{|\vec{r} - \vec{r}\,'|} e^{-\frac{i\sigma_1}{c}|\vec{r} - \vec{r}\,'|} \int_{-\infty}^{+\infty} dt' e^{i\sigma t'} S \qquad (2.72)$$

After integration, this gives

$$\rho_\sigma(\vec{r}) = \frac{1}{8\pi^2 c^3} \int \frac{d\vec{r}}{|\vec{r} - \vec{r}\,'|} \int_{-\infty}^{+\infty} dt' e^{\frac{-i(\sigma t' + \sigma_1|\vec{r} - \vec{r}\,'|)}{c}} \rho_o^{1/2}(\vec{r}) \qquad (2.73)$$
$$[\, -\sigma_1 n_i' n_j' u_i' u_j' + i2\sigma_o \sigma_1 n_j' w' u_j' + \sigma_o^2 w'^2]$$

where we have written u'_1, u'_2, w' for the components of $u(\vec{r}\,', t')$, and n'_i for the components of the unit vector $\frac{r - r'}{|\vec{r} - \vec{r}\,'|}$.

The calculation of the intensity

$$I = \langle p \vec{u} \rangle = - c^6 \langle \rho' \left(\frac{\partial}{\partial t}\right)^{-1} (\vec{\nabla} - \frac{1}{2c^2}\vec{g})\rho' \rangle \qquad (2.74)$$

is obtained with help of the Fourier transform and introduces, naturally, correlation functions. Taking into account the symmetry properties of the turbulence, Unno obtains:

$$I\,(\vec{r}) = \frac{\vec{n}}{16\pi^{3}\;r - r'^{\,3}}\int_{V} d\vec{r}\,'\rho_{o}(\vec{r}\,')\int_{o}^{\infty} d\sigma_{1}\;d\vec{\xi}\int_{\infty}^{+\infty} d\tau\;\cos(\sigma\tau)\;\frac{\sigma_{1}^{2}/c^{5}}{\sigma_{1}^{2}+\sigma_{o}^{2}}\left(f_{q}\,\sigma_{1}^{4} + f_{d}\,{}_{o}^{2}\,\sigma_{1}^{2} + \sigma_{o}^{4}\right) < w'^{2}\;w''^{2} > \tag{2.75}$$

where the correlation function $< w'^{2}\; w''^{2} >$ is taken in between two points separated by the distance $\vec{\xi}$ and the time interval τ. The factors $f_{q}\,(\vec{n})$ and $f_{d}\,(\vec{n})$ depend on the properties of the turbulence.

To illustrate the nature of the various terms contributing to $I(\vec{r})$, Unno adopts an idealized form of the correlation formation,

$$< w'^{2}\;w''^{2} > = U_{o}^{4}\exp\left[-1/2\;\frac{(r^{2}+U_{o}^{2}\,\tau^{2})}{\ell_{o}^{2}}\right] \tag{2.76}$$

where U_{o} and ℓ_{o} are the characteristic velocity and length of the turbulence. Two cases have to be considered:

(a) The size of the turbulent elements is small compared to the scale height, $\ell_{o} < 2\;(U_{o}/c)H$. Then,

$$I(\vec{r}) = \frac{\vec{n}}{4\pi R^{2}}\int_{V} d\vec{r}\,'\;\frac{\sqrt{\pi}}{2}\;\frac{\rho_{o}U_{o}^{3}}{\ell_{o}}\left[6f_{q}\left(\frac{U_{o}}{c}\right)^{5} + \frac{1}{2}f_{d}\left(\frac{U_{o}}{c}\right)^{3}\left(\frac{\ell_{o}}{H}\right)^{2} + \frac{1}{8}\left(\frac{U_{o}}{c}\right)\left(\frac{\ell_{o}}{H}\right)^{4}\right] \tag{2.77}$$

where $f_{q} \cong 5$ according to Proudman (1952). Dipole and monopole radiation are negligible at low Mach number.

(b) The size of the turbulent elements is comparable to the scale height, $\ell_{o} > 2\;(U_{o}/c)H$.

This gives:

$$I(\vec{r}) = \frac{\vec{n}}{4\pi |r - r'|^2} \int_V d\vec{r}' \frac{\sqrt{\pi}}{2} \frac{\rho_o U_o^3}{\ell_o} \left[120 f_q \left(\frac{U_o}{c}\right)^7 \left(\frac{\ell_o}{H}\right)^{-2} + 6 f_d \left(\frac{U_o}{c}\right)^5 + \frac{1}{2} \frac{U_o}{c} \left(\frac{\ell_o}{H}\right)^2 \right]$$
$$\times \exp\left[-\frac{1}{8} \left(\frac{c\ell_o}{U_o H}\right)^2 \right] \tag{2.78}$$

The dipole and monopole terms are important at low Mach number, but the acoustic power is greatly reduced by the exponential factor due essentially to the existence of a characteristic frequency which cuts down all lower frequencies.

Let us consider the intensity due to the high frequency part of the spectrum. The convergence of the integral results from the decrease of (U_o/c) with depth. As an order of magnitude, we obtain for the flux of mechanical energy (with $f_q = 5$) :

$$\Phi_{mec} \cong \frac{H}{\ell_o} \cdot \left(\frac{W_o}{c}\right)^5 \cdot \frac{2aRQ}{\mu C_p} \Phi_{conv} \tag{2.79}$$

As an example, we take for the sun $(W_o/c) = 1/4$, $(H/\ell_o) = 10$, $(2aRQ/\mu C_p) \simeq 1/2$, which gives:

$$\Phi_{mec} \simeq 3 \cdot 10^8 \text{ ergs cm}^{-2} \text{ sec}^{-1} \tag{2.80}$$

We shall see later the conclusions which can be derived from this result.

As far as the spectrum of the acoustic waves is concerned, it comes essentially from the turbulence spectrum. There is presently no theory giving the spectrum. From the observations, there are vague indications, as those given by Schwarzschild and Richardson (1950) for the sun, but which cannot be considered as very reliable (for the spectroscopic reasons given above.)

C. Waves in the Solar Atmosphere

In the case of the solar atmosphere, some information

can be obtained from spectra having a high resolving power both in space, time, and wave length.

However, the observations show the low frequency oscillations which are essentially gravity waves. In the following material, we shall consider different aspects of linear filtering by the solar atmosphere: (a) the effect of the radiative damping; (b) the spectrum as a function of height; (c) the role of the temperature minumum.

A very simple model set up by P. Souffrin (1966) shows how gravity waves can be cut by the radiative damping.

If we consider a bubble of density ρ^*, surrounded by a medium of density ρ, the equation of motion is

$$\rho^* \frac{d^2\xi}{dt^2} = -g\ (\rho^* - \rho). \tag{2.81}$$

The energy equation can be written as

$$dp^* - \frac{\gamma p^*}{\rho^*} d\rho^* - \rho^* (\gamma - 1)\ dQ = 0 \tag{2.82}$$

in which the heat exchange dQ is given by Newton's law of cooling

$$dQ = -\frac{C_v}{\tau} (T^* - T)\ dt \tag{2.83}$$

in which τ is the cooling time.

This leads to the differential equation for the velocity of the bubble:

$$\frac{d^2 v}{dt^2} + \frac{3}{2}\frac{\gamma - 1}{\gamma}\frac{1}{\tau}\frac{dv}{dt} + \omega_v^2\, v = 0 \tag{2.84}$$

where ω_v is the Väisälä frequency.

This corresponds to an oscillating motion only if

$$\tau > \frac{3}{4}\frac{\gamma - 1}{\gamma}\ \omega_v^{-1} \tag{2.85}$$

When considering the oscillatory motion in an atmosphere, all waves are damped. In the diagnostic diagram (Fig. 10), we separate propagating and non-propagating regions by the condition that the damping length is equal to the wave length. When the radiative relaxation time vanishes, the domain of the gravity waves vanishes also.

The relaxation time in an atmosphere deserves some attention. The best which can be done at the present time is to use the results of Spiegel (1957) which have been obtained for a uniform medium and give, at least, orders of magnitude.

Let us call q the energy lost by radiation per unit volume and unit time, and U the internal energy of the gas. The equation of energy reads:

$$\rho \frac{dU}{dt} + p \vec{\nabla} \cdot \vec{v} = q \tag{2.86}$$

If B is the source function, $\bar{I}$ the mean intensity, $\mathcal{H}$ the absorption coefficient per unit volume, then the emissivity q is:

$$q = 4\pi \mathcal{H} (\bar{I} - B) \tag{2.87}$$

and the expression of $\bar{I}$ is:

$$\bar{I} = \int_0^\infty \mathcal{H}(\vec{r} - \vec{r}_o)\, B(\vec{r} - \vec{r}_o)\, \frac{e^{-\tau(\vec{r} - \vec{r}_o)}}{4\pi\,(r - r_o)^2}\, d\vec{r} \tag{2.88}$$

When we perturb the gas, we have to take into account the perturbation of B, $\mathcal{H}$, and τ. If δT is the perturbation of the temperature, the equation of energy can be written to the second order:

$$\frac{d\delta T}{dt} = \frac{4\sigma \mathcal{H}}{\rho\, C_v} \left[4 \int \mathcal{H}\, T^3 \frac{e^{-\tau}}{4\pi (r - r_o)^2}\, \delta T d\vec{r} - 4T^3 \delta T + \int \mathcal{H}_{T_o} \delta T - \mathcal{H} \int \mathcal{H}_{T_o} \delta T d\vec{r}\ )\ T^4 \frac{e^{-\tau}}{4\pi (r - r_o)^2}\, d\vec{r} \right] \tag{2.89}$$

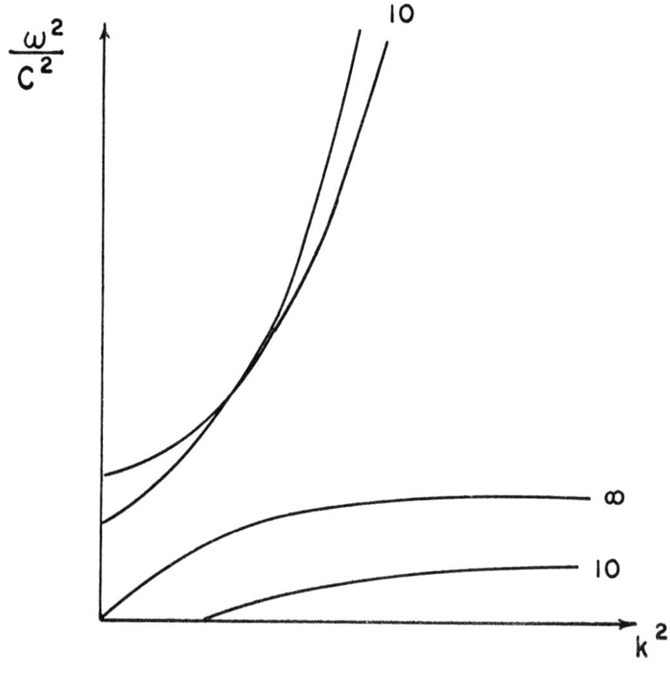

Figure 10

For an infinite, homogeneous medium, we have:

$$\frac{\partial \delta T}{\partial t} = \gamma \left[\int \delta T \frac{e^{-\mathcal{H}(r-r_o)}}{4\pi(r-r_o)^2} \mathcal{H} d\vec{r} - \delta T + \frac{\mathcal{H}_T \delta T}{4} \int (\delta T - \mathcal{H} \int \delta T dr') \frac{e^{-\mathcal{H}(r-r_o)}}{4\pi(r-r_o)^2} d\vec{r} \right] \tag{2.90}$$

where

$$\gamma = \frac{16\mathcal{H}\sigma T^3}{\rho C_v} \tag{2.90'}$$

If we look for a separable solution of the form

$$e^{nt} e^{i\vec{k}\cdot\vec{r}}$$

we can obtain for n the expression

$$n = \gamma\left(1 - \frac{\mathcal{H}}{k} \operatorname{Cot}^{-1} \frac{\mathcal{H}}{k}\right) \tag{2.91}$$

If the optical thickness is small, $(\mathcal{H}/k) \to 0$, we can obtain

$$n = \gamma \tag{2.91'}$$

If the optical thickness is large, $(\mathcal{H}/k) \to \infty$ and $n \to k^2(\gamma/3\mathcal{H}^2)$.

The relaxation time in the solar atmosphere is given by Noyes and Leighton, using the Spiegel expression for the optically thin case:

$$\tau = \frac{C_v}{16\bar{\mathcal{H}}\sigma T^3} \tag{2.92}$$

where $\bar{\mathcal{H}}$ is now the absorption coefficient per unit mass, and is reproduced in Fig. 11.

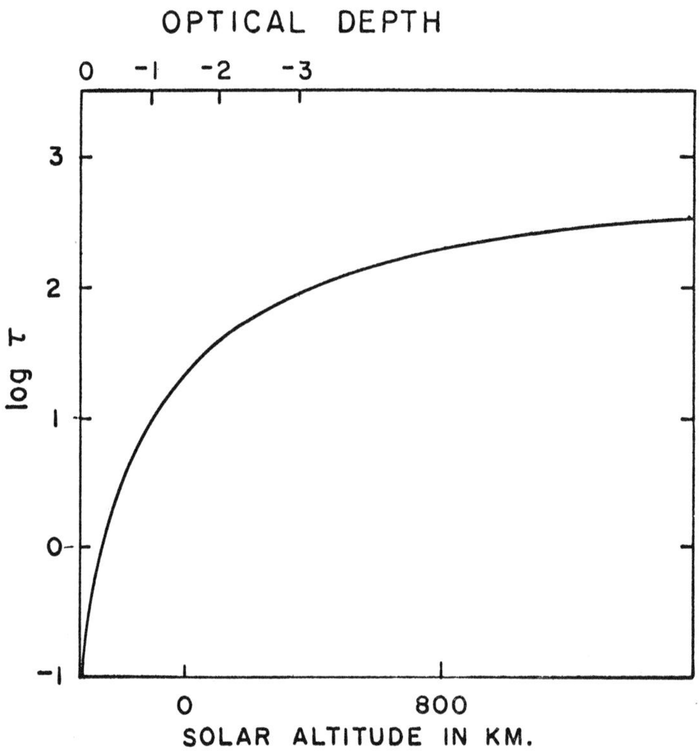

Figure 11

We can see from the figure that low frequency waves cannot propagate in the upper layers of the photosphere, at least, up to an optical depth of 10^{-3} . This suggests the idea that the atmosphere behaves like a filter which damps the low frequencies.

Generally speaking, we can consider at a certain level an incident wave $\bar{I}(\omega, \vec{k})$, a reflected wave $R(\omega,\vec{k})$ and, higher in the atmosphere, a transmitted wave $T(\omega,\vec{k})$, and we have a linear relation

$$T(\omega,\vec{k}) = f(\omega,\vec{k})\ \bar{I}(\omega,\vec{k}) \tag{2.93}$$

Let us consider an atmosphere with a temperature minimum.

If we consider a wave which propagates upwards, it experiences both reflection and transmission by the temperature minimum. If we consider the oscillations of the atmosphere, with only outgoing waves from the temperature minimum, the atmosphere experiences acoustic damping, and for k real, ω is complex. The values of ω are given by the poles of f. The poles of f actually correspond to the free oscillations of the atmosphere, with no incident wave from below, and equation (93) should read:

$$\bar{I}(\omega,\vec{k}) = \frac{1}{f(\omega,\vec{k})}\ T(\omega,\vec{k}) = 0 \tag{2.94}$$

Although this effect is important in the sense that the atmosphere will select those complex eigenfrequencies, it is not very important physically since the pole in ω is usually fairly far from the real axis and the atmosphere selects, in fact, a whole spectrum; the further the pole is from the real axis, the flatter the spectrum.

As emphasized in a paper by Schatzman and Souffrin (1967), a difficulty is that when the temperature profile is adjusted to give the pseudo period observed by Noyes and Leighton, and by Evans, Michard, and Servajan, the damping is too slow; this discrepancy is not likely to be removed by taking into account horizontal motions because this will merely increase the reflectance, then decrease the leakage and the damping rate also.

This can be shown quantitatively in the study of the behavior of a three-layer model without gravity, including oblique waves (Schatzman, 1964).

If we consider the analysis of the observations by Mein (1966), it seems very likely that the change of the power spectrum with altitude is essentially due to the linear filtering. A semi-qualitative explanation can be found in the case of the isothermal atmosphere. Starting at level z_o with a Fourier component $F_o(k\omega)$ of the spectrum, the power spectrum at altitude z is given by:

$$\underline{P}_z(\omega) = \int dk\, k\, F_o(k,\omega)\, e^{-2\lambda z} \tag{2.95}$$

with

$$\lambda = \left[k^2 - \frac{\gamma\omega^2}{c^2} - \frac{1}{4H^2}\right]^{1/2} \quad \text{if} \quad \frac{\gamma\omega^2}{c^2} - \frac{1}{4H^2} < k^2$$

$$\lambda = 0 \quad \text{if} \quad \frac{\gamma\omega^2}{c^2} - \frac{1}{4H^2} > k^2$$

The effect of the filtering has been calculated for a source defined by the auto-correlation function

$$\Phi(r,\tau) \sim \exp\left[-\left(\frac{\tau}{\theta} + \frac{r^2}{2L^2}\right)\right] \tag{2.96}$$

P. Mein (1966) has compared the observed power spectrum $f(\omega)$ with the theoretical one. The theoretical spectrum has been calculated including the radiative damping of the waves. In the range 240-300 seconds, the agreement is satisfactory, and the power spectrum can be explained by evanescent waves. The intensity of the high frequency waves is roughly proportional to $\rho^{-1/2}$ as it would be expected for acoustic waves.

D. Wave Propagation and Shock Waves.

The energy flux which goes into the higher atmospheric layers is carried by mechanical waves. It is easy to show, by considering the integral over a finite volume, that the energy flux for a plane horizontal wave is:

$$F_{mech} = \frac{1}{2} \rho v^2 V_G \tag{2.97}$$

in which the group velocity is given by

$$V_G^2 = c^2 \left(1 - \frac{\gamma^2 g^2}{4c^2 \omega^2}\right) = c^2 \left(1 - \frac{\omega_c^2}{\omega^2}\right) . \tag{2.98}$$

It is quite clear that the flux of mechanical energy vanishes for $\omega < \omega_c$.

This should naturally be supplemented by the results concerning the properties of oblique waves.

When there is no damping in an isothermal atmosphere, the group velocity is constant, and the velocity v increases with height, as $\rho^{-1/2}$.

Let us now consider the case of radiative damping. For simplification, we shall take horizontal waves, with $\omega_R \gg 1$. In that case, the amplitude is something like:

$$\exp \left[i(\omega t - \frac{\omega}{c} z) - \frac{\gamma - 1}{2c\tau_R} z + \frac{z}{2H} \right] \tag{2.99}$$

As can be expected, the damping limits the amplitude growth. To that approximation, the amplitude remains constant when

$$\frac{2}{\gamma(\gamma - 1)} \omega_c \tau_R = 1. \tag{2.100}$$

This situation is reached for $\tau_R \simeq 25$ seconds, that is for an optical depth $\tau_{5000} \simeq 10^{-2}$.

Let us now consider the way in which compression waves are being transformed into shock waves when going up in the atmosphere.

It is easily shown in gas dynamics that a perturbation with material velocity u propagates with velocity u + c. As a consequence, the crest of the wave catches up with the trough, the difference in velocity being 2u. After propagation, we can say that the wave has become a shock when $\int 2u\,dt = \int 2u \frac{dz}{c}$ is of the order of a half wave length. In the present case, this gives:

$$\int_0^z \frac{2u_o}{c} \exp\left[z\left(\frac{1}{2H} - \frac{\gamma - 1}{2c\tau_R}\right)\right] dz \simeq \frac{\pi c}{\omega}. \tag{2.101}$$

A condition for the acoustic wave to become a shock is naturally that the damping be not too strong. This condition can be expressed as

$$\frac{\gamma(\gamma - 1)}{2\omega_o \tau_R} < 1 + \frac{4u_o \omega}{\gamma \pi g} \tag{2.102}$$

To have an idea of when it occurs, let us take $u_o = 3 \times 10^4$ km/sec., and $\omega = 1$ sec^{-1}. We obtain then

$$\begin{gathered} \frac{2}{\gamma(\gamma - 1)} \omega_c \tau_R > 0.55 \\ \omega_c \tau_R \gtrsim 0.3 \\ \tau_R \gtrsim 8 \text{ sec} \end{gathered} \tag{2.103}$$

This is satisfied in all regions above $\tau = 0.3$ and we can consider that radiation damping is not sufficient in the high layers of the photosphere to prevent acoustic waves from becoming shock waves.

In the case of constant amplitude, the distance over which an acoustic wave becomes a shock is

$$z = \frac{\pi c^2}{u_o \omega} \quad . \tag{2.104}$$

For $u_o = 3 \times 10^4$ km/sec., $\omega \simeq 1$/sec., we obtain $z \simeq 300$ KM.

We can expect the acoustic waves to become shock waves in something like 600 to 800 km above the photospheric level. Due to the fact that an irreversible process takes place through the shock front, the shock wave represents the most efficient way of transforming mechanical energy into heat.

As it is well known, the amount of heat available can be estimated from the change of the fluid's enthalpy. The enthalpy is defined as the sum of the internal energy U and of the product PV:

$$H = U + PV \tag{2.105}$$

The specific enthalpy, h, is defined for one gram of gas:

$$h = e + P\tau = e + \frac{P}{\rho} = \frac{\gamma}{\gamma - 1} \frac{P}{\rho} \tag{2.106}$$

where τ is the specific volume of the gas.

The enthalpy is important as it is a function only of the pressure P and the density ρ only.

Physically, change of enthalpy at constant pressure is equal to the amount of heat exchanged with the surrounding system.

Across the shock front, the change of specific enthalpy can be obtained by eliminating the velocities in between the equations of conservation of mass, momentum, and energy written:

$$\rho_1 u_1 = \rho_o u_o \tag{2.107}$$

$$P_1 + \rho_1 u_1^2 = P_o + \rho_o u_o^2 \tag{2.108}$$

$$h_1 + \frac{1}{2} u_1^2 = h_o + \frac{1}{2} u_o^2 \tag{2.109}$$

It is easily found that

$$h_1 - h_o = \frac{1}{2} (P_1 - P_o) \left(\frac{1}{\rho_o} + \frac{1}{\rho_1}\right) . \tag{2.110}$$

In a gas, when there is a discontinuity of density, temperature, pressure, one has to express across the discontinuity the following equations of conservation:

- conservation of mass
- conservation of momentum
- conservation of energy. This last equation is known as the Rankine-Hugoniot relation.

If we know the density, pressure, and velocity on one side of the shock front, the velocity of the shock front determines entirely the physical conditions on the other side of the discontinuity.

We can express the relations between the two sides of the shock in different ways. For example, assuming that the gas ahead of the shock front is at rest, and calling a_o the sound velocity ahead of the shock, one obtains the following relations:

$$p = \rho_o a_o^2 \left[\frac{2}{\gamma + 1} M^2 - \frac{\gamma - 1}{\gamma(\gamma + 1)}\right] = \frac{\rho_o a_o^2}{\gamma} \Phi \tag{2.111}$$

$$\rho = \rho_o \frac{(\gamma+1)M^2}{(\gamma-1)M^2 + 2} \tag{2.112}$$

$$u = a_o \frac{2}{\gamma+1} (M - \frac{1}{M}) = \xi a_o \tag{2.113}$$

where M is the Mach number of the shock velocity, $U=a_o M$ and γ the adiabatic compressibility of the gas. We shall consider, in the following, only perfect gases, for which $\gamma = \frac{5}{3}$.

We are concerned at the way in which the energy of the shock is used for the heating of the gas. This is probably one of the main problems of the shock propagation. The fluid experiences through the shock front an irreversible process; after the shock we shall suppose that the gas comes back to its initial states, but we do not know along which path the gas returns to its initial states.

We can assume (1) that the gas expands adiabatically until it comes back to the initial pressure, and then cools down until it obtains the initial density; (2) that the gas, after the shock front cools at constant pressure until it comes back to the initial adiabatics, then expands adiabatically to the initial pressure and density.

The energy available can be obtained by considering the change of enthalpy of the gas.

Along the first path, the change of enthalpy is: (Schatzman's path):

$$h = \frac{P - P_o}{2}\left(\frac{1}{\rho} + \frac{1}{\rho_o}\right) - \frac{\gamma}{\gamma - 1}\left(\frac{P}{\rho} - \frac{P_o}{\rho_o}\right) \tag{2.114}$$

with

$$\frac{P}{P_o} = (\rho/\rho*)^{\gamma} \tag{2.115}$$

Along the second path, the change of enthalpy is (Weymann's path):

$$h = \frac{1}{\gamma - 1}\frac{P_o}{\rho_o}\left\{\left(1 - \frac{\pi}{2\gamma}\right) - \left(1 - \frac{\pi}{2}\right)\left(1 + \frac{\pi}{2\gamma}\right)^{\gamma}\left(1 - \frac{\pi}{2\gamma}\right)^{-\gamma}\right\} \tag{2.116}$$

$$\text{with } \pi = \frac{P - P_o}{P_o} \tag{2.117}$$

When the shock strength is small (π small compared to unity) these two expressions reduce to the same form:

$$h = \frac{\gamma + 1}{12\,\gamma^2}\,\frac{P_o}{\rho_o}\left(\frac{P - P_o}{P_o}\right)^3 = \frac{2}{3}\,a_o^2\,\frac{(M^2 - 1)^3}{(\gamma + 1)^2} \tag{2.118}$$

This can be written:

$$\rho\,\Delta h = \rho\,\frac{\gamma + 1}{12}\,\frac{v^3}{c}\;. \tag{2.119}$$

We shall accept presently the idea that the effect of the pressure and density gradients are negligible. Therefore, we use expression (2.119) for the energy available and try to work out an equation of transport for the mechanical energy.

A sine wave would transform into an "N" wave, with a constant interval of time between the shocks equal to the period. The mechanical energy which is available in the wake of an "N" wave is of the order of $1/6\ v^2\rho c\ \tau$, and this gives the rate of dissipation of "N" waves:

$$\frac{d}{dx}\,\frac{1}{6}\,\rho c \tau v^2 = -\,\frac{\gamma + 1}{12}\,\frac{\rho v^3}{c}\;. \tag{2.120}$$

The flux of mechanical energy is obtained by multiplying the energy of a wave by its frequency $\frac{1}{\tau}$ to give

$$\frac{d}{dx}\,\frac{1}{6}\,\rho c v^2 = -\,\frac{\gamma + 1}{12}\,\frac{\rho v^3}{c\tau}\;. \tag{2.121}$$

This expression cannot be used, for we know that acoustic waves, when propagating in a region of varying temperature, will refract.

If I is the intensity of the mechanical energy which is transported in a certain direction, $\frac{I}{n^2}$ is conserved during propagation. Following a ray path, this gives for no dissipation:

$$\mu\,\frac{\partial I}{\partial z} - \frac{1}{c}\,\frac{\partial c}{\partial z}\,\frac{\partial}{\partial \mu}\,I\,(1 - \mu^2) = 0, \tag{2.122a}$$

and with dissipation:

$$\mu \frac{\partial I}{\partial z} - \frac{1}{c} \frac{\partial c}{\partial z} \frac{\partial}{\partial \mu} I (1 - \mu^2) = - \frac{\gamma + 1}{12} \frac{\rho v^3}{c\tau} . \quad (2.122b)$$

As it is not obvious that there is a relation between the dissipation and the intensity in expression (2.122b), we shall write the non-linear dissipation term

$$\frac{\gamma + 1}{12} \frac{\rho v^3}{c\tau} = \frac{\gamma + 1}{2} \frac{Iv}{c^2 \tau} . \quad (2.123)$$

We can then multiply by $2\pi d\mu$ and integrate $\mu = \cos\theta$ from -1 to +1 to obtain:

$$\frac{dF}{dz} = - \frac{\gamma + 1}{2} \frac{v}{c} \frac{1}{c\tau} 4\pi \bar{I} . \quad (2.124)$$

In the same way, we can multiply (2.122a) by $(\frac{1}{2})$ $\mu d\mu$ and obtain:

$$\frac{4\pi}{3c^2} \frac{\partial}{\partial z} \bar{I} c^2 = - \frac{\gamma + 1}{2} \frac{v}{c} \frac{F}{c\tau} \quad (2.125)$$

using the Eddington approximation.

We see clearly that the mechanical flux will be constant in the absence of dissipation. In that case the waves are refracted downwards with the flux of up-going waves exactly equal to the flux of down-going waves, and the total flux is zero.

Coming back to the expression of I as a function of v, we obtain:

$$\frac{dF}{dz} = - (\gamma + 1) \frac{\pi}{3} \frac{\rho v^3}{c\tau} , \quad (2.126)$$

and

$$\frac{1}{c^2} \frac{d}{dz} \frac{1}{6} \rho c^3 v^2 = - \frac{3}{8\pi} \frac{\gamma + 1}{2} \frac{v}{c} \frac{1}{c\tau} F. \quad (2.127)$$

It is very important to notice that the set of equations (2.126) and (2.127) associated with the equation of conservation of energy, i.e., radiative flux, conductive flux, and mechanical flux,

$$\frac{dF}{dz} + \frac{d}{dz}(F_{rad} + F_{cond}) = 0, \tag{2.128}$$

gives the possibility of explaining the sharp rise in temperature at the boundary of the chromosphere.

However, the difficulty remains that the decrease of the mechanical flux due to refraction brings little energy in the high layers of the chromosphere. To show this effect, we can take, as a first approximation, zero dissipation which gives $\bar{I}c^2$ = constant, or $\rho v^2 c^3$ = constant for the mean intensity of acoustic waves, from which we derive

$$v^2 \sim (\rho c^3)^{-1}. \tag{2.129}$$

For div F, this gives, according to equation (2.126),

$$\text{div } F \sim \rho^{-(\frac{1}{2})} c^{-(11/2)} . \tag{2.130}$$

As the pressure varies very slowly in the region of the temperature rise, this gives div $F \sim c^{-(9/2)}$. A temperature jump from 10^4 to 10^6 °K gives a decrease of div F by a factor $10^{-4.5}$, in agreement with the results of Osterbrock (1961).

The fast rise in temperature can be attributed essentially to the refraction of acoustic waves. In the corona, the emissivity is proportional to $\rho^2 T^{-(1/2)}$ (free-bound transitions). At constant pressure, this is proportional to $T^{-(5/2)}$, which is almost exactly the same law as the law obtained for div F, which is $T^{-(9/4)}$. It can be said that a fast decrease in dissipation can be compensated for only by a fast decrease in density. For a quasi-constant pressure, this means a fast increase in temperature.

However, although the steep rise of temperature from the chromosphere to the corona can be interpreted in terms of mechanical heating, there is no quantitative agreement between the observations and the calculation. It should be especially mentioned that the amount of energy which reaches the corona is too small to explain the amount of energy which is radiated away. Therefore, another mechanism has to be introduced to carry energy.

E. Convective Zone and Magnetic Field.

The observation of the surface of the sun in the K line of Ca II shows a chromospheric structure{chromospheric network or supergranulation, Leighton (1964)} with a long lifetime (50,000) seconds). The supergranulation velocity field has been shown to be responsible for the well-known reticulated structure of the chromospheric network (Simon and Leighton, 1964). A key feature of this relationship is that the cellular circulatory motion concentrates the weak, large-scale field of a magnetic region into narrow lanes at the boundary of the convection cells. Suggested by Pikelner (1962), the calculation of the magnetic field produced by convection has been made by Parker (1963). The equation for the evolution of the magnetic field

$$\frac{\partial \vec{B}}{\partial t} = \nabla \times (\vec{v} \times \vec{B}) + \lambda \nabla^2 \vec{B} \tag{2.131}$$

can be solved for a given motion. If the diffusion term is negligible (see below, the dynamo problem), it is especially simple to integrate, as it is a linear differential equation for $\vec{B}$.

Parker considers perfectly conductive liquid with lines of flow similar to convective flow, with

$$v_x = \frac{\partial}{\partial y} \Phi \qquad v_y = \frac{\partial \Phi}{\partial x} \qquad \text{and} \quad \Phi = vy \sin kx \tag{2.132}$$

A vertical component of the magnetic field is dragged against the edge of the convective pattern and increases like exp(kvt).

This is similar to the result of Spitzer (1957) and of Weiss (1964) that convective motion ejects the magnetic field from the vortex and concentrates it in the boundary of the convective cell.

The succession of the convective cells is at random. Leighton (1964) has shown that supergranulation convective currents produce dispersal and migration of the magnetic field by a random walk process.

The convective cells of the supergranulation are probably related (as suggested by Böhm) to the convection in all the depth of the convective zone.

The supergranulation is also visible in the chromosphere (chromospheric network). The chromosphere is known to be non-homogeneous, and this is shown first of all by the pictures in the red light of Hα that show clearly thin spikes of bright matter called spicules, which are more dense than the surrounding material. The interpretation of the chromospheric spectrum requires the assumption of a two-temperature model. In other words, there are regions with a temperature $T_1(z)$ and regions with another temperature $T_2(z)$.

We have been considering before the propagation of waves in a plane parallel atmosphere, but this is naturally a first approximation. The propagation takes place in a non heterogeneous medium. To show that the heterogeneities of the chromosphere are not a small effect which can be easily neglected, let us compare characteristic dimensions of the heterogeneities with the size of a wave front when reaching the chromosphere from a point source. To that effect let us consider a velocity of propagation varying as exp $(\frac{z}{H'})$, where H' is the vertical scale of the growth of the velocity.

The refraction of the rays for that simple law gives the following paths:

$$z = H' \log \left\{ \frac{\sin\left(\frac{x}{H'} + \Phi_o\right)}{\sin \Phi_o} \right\} ; \qquad (2.133)$$

it is easily seen that the full extension of the ray is of the order of $2\pi H'$. With a scale height of the temperature growth of the order of 2000 km, this gives a horizontal scale of the order of 12,000 km.

When the surface of the sun is observed in the light of the K line of Ca II, one discovers a pattern which has a characteristic size of about 30,000 km. This pattern is not a permanent feature of the sun, but is continuously changing, being replaced by a new pattern with a lifetime of about 16 hours. The observations of the radial velocities show that matter flows away from the center of the cells of supergranulation with a speed of about 0.3 km/sec. This can clearly be understood as due to the motion of big convective cells in the convective zone (Fig. 12a). The observations also show that the spicules are distributed like spikes along the lines of the supergranulation, and that the magnetic field is also stronger along the boundary of the supergranules (Fig. 12b). This confirms the idea of Pikelner (1962) and Parker (1963).

There is certainly a different type of thermal equilibrium in these regions of higher magnetic field, due to a difference in the amount of energy dissipated into heat by the mechanical waves. For the time being, we are interested in noting that the size of the wave front produced by a point source of noise is not negligible compared to the size of the supergranulation. There is no doubt that we must consider the diffraction and refraction of the acoustic waves by the heterogeneities.

The question arises: how large is the magnetic field in the heterogeneities of the sun? Severny (1965) has attacked

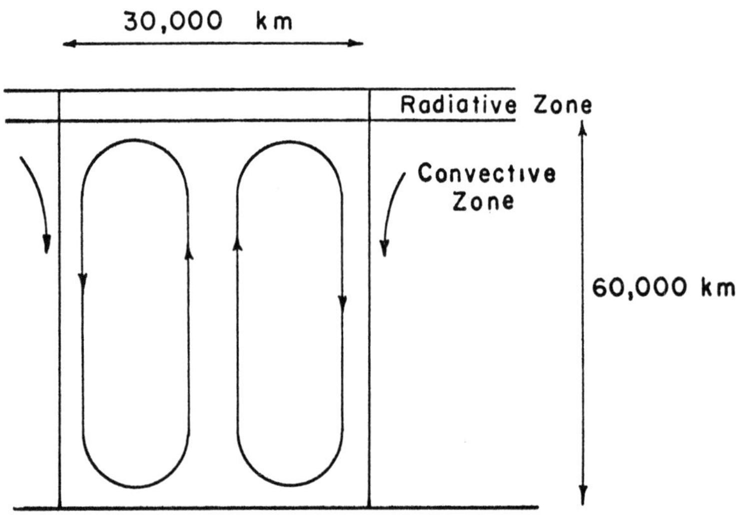

Figure 12a

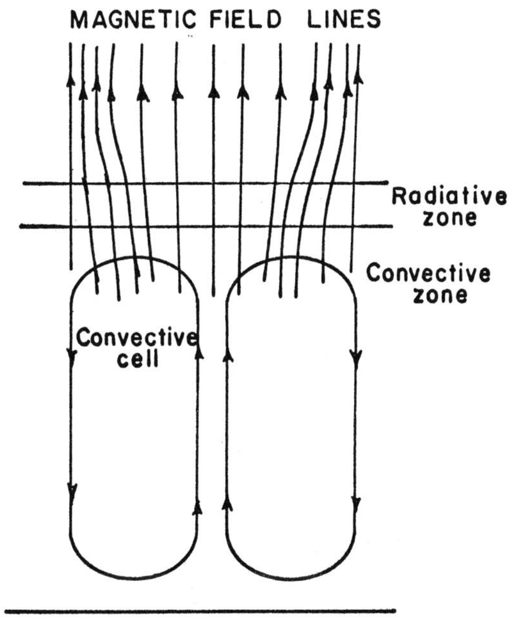
MAGNETIC FIELD LINES
Radiative zone
Convective zone
Convective cell

Figure 12b

the problem to find a relation of the maximum value of the magnetic field with the resolving power of the instrument. He finds that the higher the resolving power, the larger is the maximum value of the field; the maximum value of the magnetic field for a resolving power of 2" is of the order of 30 gauss. It is quite possible that the smallest magnetic spots on the surface of the sun correspond to a magnetic field of 100 to 200 gauss.

It is quite certain that it is not legitimate to study the wave propagation in a chromospheric model of plane parallel layers. As an example we can quote Uchida (1967), who studied the problems of acoustic waves trapped inside a cylindrical wall on which the dilation vanishes. The cylindrical wall is taken with a radius of 30,000 kilometers, which is of the order of magnitude of the supergranulation cell's magnitude.

F. Spicules.

The spicules seem to correspond well to regions of a higher magnetic field, along the edges of the supergranulation.

It seems quite probable that the heat balance in the regions of a higher magnetic field is different from what it is in regions of a smaller magnetic field.

The mechanism of formation of the spicules would then be the following: because of a change in the heat balance, matter would begin to cool and to increase in density. This would be accompanied by a mass motion, due to the fact that the matter would be unable to radiate away all the mechanical energy which it is receiving, as it is described by Kuperus and Athay (1967) and by Skalafuris (1968).

Mass motion would take place in a magnetic field which has the structure of a force-free field. The lines of force being twisted, the upward motion of the matter will be accompanied by a vortex motion which then can explain the observed rotation of the spicules.

G. Heating of the Corona. Alfvén Waves.

The amount of energy fed into the corona is fairly large. If we include both the energy radiated away and the energy transported by thermal conductivity both inwards to the chromosphere and outwards to interplanetary space, we find, according to Billings (1966), a flux of the order of 10^6 ergs cm^{-2} sec^{-1} [a recent critical study of Athay (1967) gives one order of magnitude less] .

If we take into account the refraction effect, we find no way of bringing that amount of energy to the corona by acoustic waves. This is why it has been suggested that Alfvén waves are the heating agent of the corona. A good reason for considering the effect of Alfvén waves is that the damping of Alfvén waves at the coronal level is very small and, consequently, they can carry energy very far.

Dissipation of Alfvén waves can be due both to Joule heating and friction. Piddington (1956) has shown that the friction on the neutrals of the charge particles carried by the lines of force dissipates energy. Osterbrock (1961) has estimated both effects.

The conclusion of Osterbrock is that the scale heights for damping Alfvén waves is very small at the photospheric level, and, therefore, Alfvén waves cannot be generated at photospheric levels (Table IX)

Table IX

Damping Length of Alfvén Waves

H (km)	(km) B = 2 gauss	B = 50 gauss
-400	7.4×10^{-2}	1.2×10^3
-300	1.4×10^{-2}	2.2×10^2
-200	2.7×10^{-2}	4.2×10^2
-100	1.1×10^{-1}	1.8×10^3
0	2.6	8.2×10^3
+250	1.1×10^2	1.2×10^5
+500	2.1×10^3	1.7×10^6

The conclusion is that Alfvén waves, and essentially the slow mode, are generated at the chromospheric level. We shall then consider the problem of the generation of Alfvén waves through a coupling with the acoustic mode. To approach the problem, we first review basic properties of wave propagation in a plasma in the presence of a magnetic field.

It is well known that, in the presence of a magnetic field, three modes propagate. If we consider the linearized equation of motion in a uniform fluid, the three modes are entirely decoupled.

If A is the Alfvén velocity, then $A^2 = \frac{1}{\mu_0}\frac{B^2}{\rho}$, and a the sound velocity $a^2 = \frac{\gamma P}{\rho}$, and the equation of dispersion gives for the velocity C,

$$(C^2 - A^2 \cos^2 \theta)\ (C^4 - C^2 (a^2 + A^2) + a^2 A^2 \cos^2 \theta) = 0, \tag{2.134}$$

in which θ is the angle between the magnetic field and the direction of propagation. The velocity surfaces and the surfaces of indices of refraction are surfaces of revolution around the magnetic field direction (Fig. 13).

It is important to consider the meaning of these surfaces by examining the direction of the vibrations, i.e., the polarizations of the waves.

The Alfvén wave corresponds to a vibration which is both perpendicular to the magnetic field and to the direction of propagation; i.e., an oscillation in which the restoring force comes from the curvature of the lines of force.

The other two modes are a slow mode and a fast mode. When the sound velocity is larger than the Alfvén velocity, the vibration of the fast mode is almost along the propagation direction (Fig. 14 a). It is very much like a sound wave distorted by the presence of the magnetic field. The vibration of the slow mode is almost perpendicular to the direction of the propagation and is very similar to an Alfvén wave distorted by the pressure effects.

When the Alfvén velocity is larger than the sound velocity, the vibration of the fast mode is almost perpendicular to the magnetic field, and is a type of Alfvén wave distorted by pressure effects (Fig. 14 b). The slow mode is almost along the magnetic

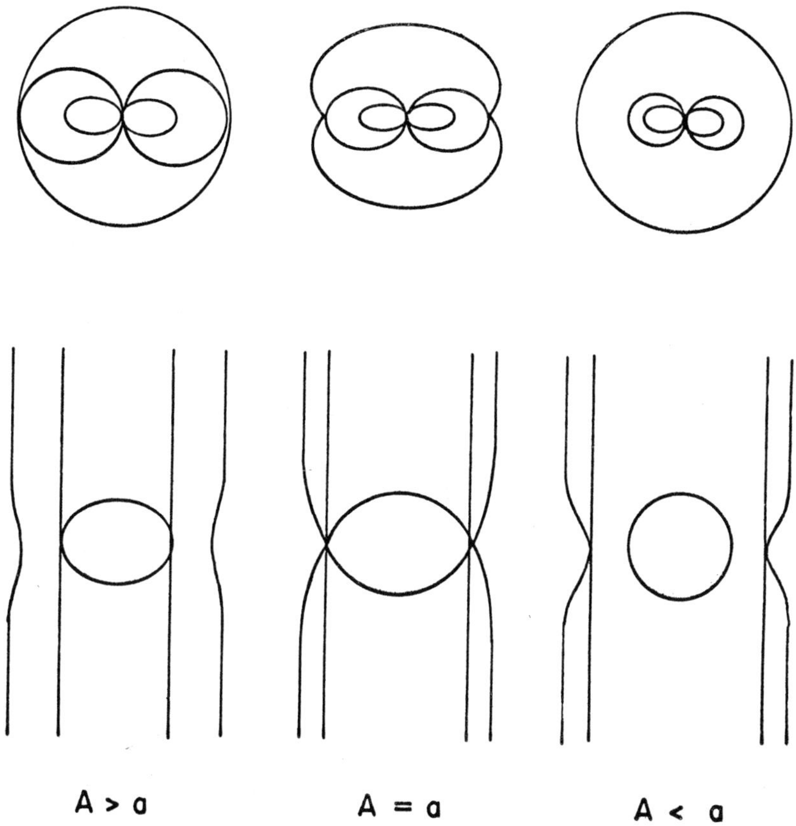

Figure 13

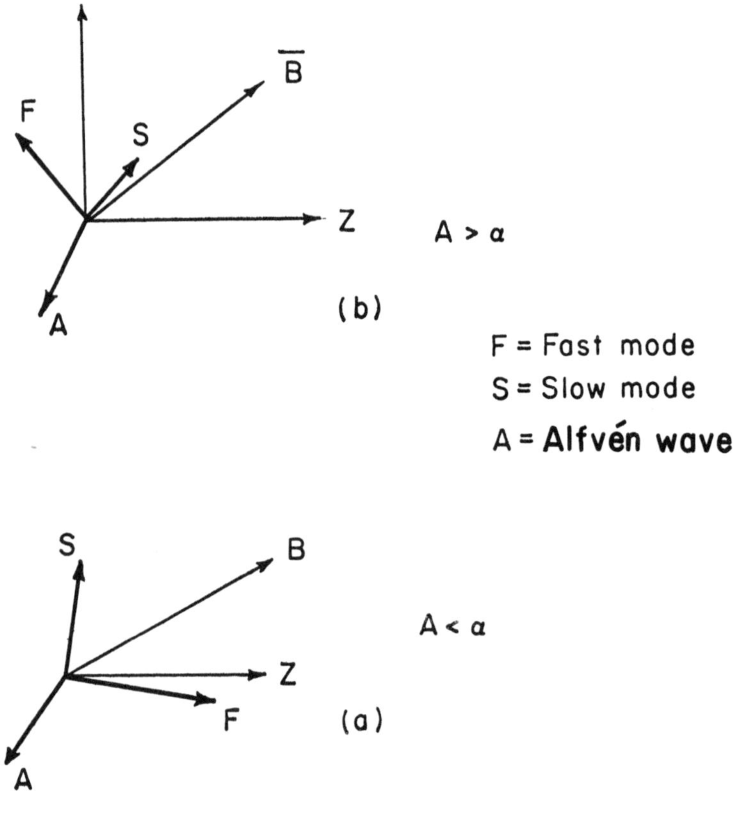

Figure 14

field, and is a type of acoustic wave distorted by the presence of the magnetic field. The three vectors are perpendicular to each other. In other words the three directions of polarization are orthogonal to each other.

The problem of refraction of a wave at a discontinuity has been considered by many authors. Generally speaking, when one mode is incident on a discontinuity, the two other modes can be generated by reflection and refraction. This is not always the case, as can be seen by considering the polarization of the wave. For example, if the magnetic field is continuous through the discontinuity, the Alfvén mode and the magneto-acoustic mode remain decoupled. Generally speaking, the coupling can take place through the non-linear terms or through the gradients of density, pressure, and magnetic field.

The general case of the propagation of waves in a non-homogeneous medium has not been considered. What we want to consider now are two examples of how Alfvén waves can be generated.

H. Waves in a Stratified Medium.

Let us consider a horizontal plane wave moving in a stratified medium. In such a medium all quantities depend only on the z coordinate. This is the problem considered by Frisch (1964).

Let us consider the components δu_x, δu_y, δu_z, of the velocity, with

$$X = \begin{Bmatrix} \delta u_x \\ \delta u_y \\ \delta u_z \end{Bmatrix} \tag{2.135}$$

It is possible to express the propagation in the following form:

$$\omega^2 X = AX'' + BX' + CX; \tag{2.136}$$

(the "primes" expressing derivation with respect to z), which can be obtained from the linearized equation of motion. The expression of the matrices $\|A\|$, $\|B\|$, and $\|C\|$ is the following:

$$
A = \left\| \begin{matrix} -B_z^2 & 0 & B_z B_x \\ 0 & -B_z^2 & B_z B_y \\ B_z B_x & B_z B_y & -(B_x^2 + B_y^2 + a^2 \mu_o \rho) \end{matrix} \right\| \tag{2.137}
$$

$$
B = \left\| \begin{matrix} 0 & 0 & \dfrac{2\,\partial B_x}{\partial z} \dfrac{B_z}{\mu_o \rho} \\ 0 & 0 & \dfrac{2\,B_z}{\mu_o \rho} \dfrac{\partial B_y}{\partial z} \\ \dfrac{B_z}{\mu_o \rho} \dfrac{\partial B_x}{\partial z} & \dfrac{B_z}{\mu_o \rho} \dfrac{\partial B_y}{\partial z} & -\left\{ \dfrac{1}{\rho} \dfrac{\partial}{\partial z} \left(\dfrac{3}{2} \dfrac{B_t^2}{\mu_o} + u^2 \rho + P \right) \right\} - \dfrac{\partial \Phi}{\partial z} \end{matrix} \right\|
$$

$$
C = \left\| \begin{matrix} 0 & 0 & \dfrac{B_z}{\mu_o \rho} \dfrac{\partial^2 B_x}{\partial z^2} \\ 0 & 0 & \dfrac{B_z}{\mu_o \rho} \dfrac{\partial^2 B_y}{\partial z^2} \\ 0 & 0 & -\dfrac{1}{\rho} \left\{ \dfrac{1}{2\mu_o} \dfrac{\partial^2 B_t^2}{\partial z^2} + \dfrac{\partial^2 \Phi}{\partial y^2} + \dfrac{\partial \Phi}{\partial z} \dfrac{\partial \rho}{\partial z} \right\} \end{matrix} \right\|
$$

Matrices B and C express the coupling of the waves through the gradients, whereas A expresses simply the propagation in a uniform medium. Matrix C contains only second-order derivatives or products of first-order derivatives with respect to z. It can be considered small compared to B.

Therefore, matrix B is the only one which will express the coupling properties of the modes through the gradients. There is no exact solution of equation (2.136) of the simplified problem.

The result is the following. Consider the direction of the magnetic field given by its asimuth ψ and its colatitude θ (Fig. 15). If the azimuth ψ does not change, the Alfvén mode

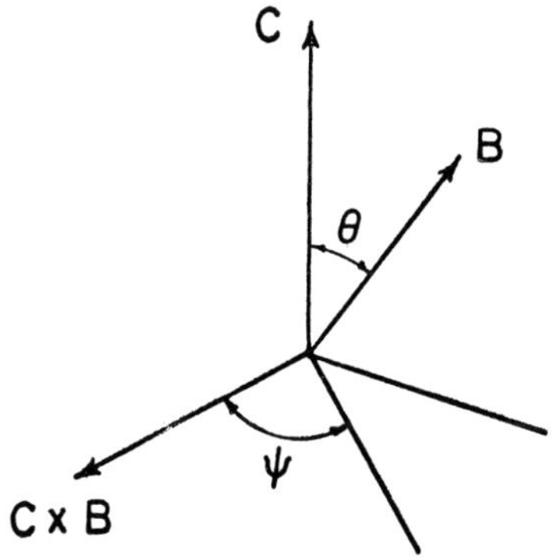

Figure 15

and the magnetosonic modes remain uncoupled. The change in colatitude θ produces a coupling between the fast and the slow mode. If the azimuth ψ changes, it produces a coupling between the Alfvén mode and the fast mode, and a coupling between the Alfvén mode and the slow mode.

The amount of coupling can be estimated by considering the amplitude of mode 2, generated between altitudes 0 and z by mode 1, with a coupling coefficient C_{12},

$$m_2(z) = \int_0^z C_{12}(\eta) \left[\exp \int_0^k js(t)dt\right] d\eta \qquad (2.138)$$

with

$$js(z) = j \left|(k_1(z) - k_2(z)\right| \quad (j = \sqrt{-1}) \qquad (2.139)$$

in which k_1 and k_2 are the wave numbers of the modes 1 and 2 for the frequency ω.

If we consider the coupling taking place over a small distance, the exponential term can be suppressed, leaving the effect of the coupling coefficient. However, it is quite clear that mode 1 will continuously feed mode 2, but this feeding will take place out of phase due to the difference in the velocity of propagation of mode 1 and 2. A beat effect will take place, similar to Landau damping, and this will limit the growth of mode 2. In a domain where k_1, k_2, and C_{12} can be taken as constants, this gives:

$$|m_2(z)|^2 = \frac{4}{s^2} C_{12}^2 \sin^2 \frac{sz}{2} \qquad (2.140)$$

for the square of the amplitude. The coupling constant C_{12} for the coupling between the Alfvén and the fast mode is of the order of $\frac{d\psi}{dz}$.

Severny (1965 b) has determined the direction of the magnetic field from different parts of the spectral line λ5250.

Since the different parts of the line form at different altitudes, this result gives an idea of the change of the magnetic field with height. Whereas the effect is very strong near sunspots, it still exists in regions of weaker magnetic fields. As an example, if $sz \ll 1$ in the coupling region, and if $(d\psi/dz) \simeq (\pi/2H_m,)$ with a scale height H_m for the rotation of the magnetic field of the order of 1000 km, the coupling is of the order of 1.

I. Waves in a Random Medium.

The microstructure of the solar magnetic field seems to be so small that, instead of considering the coupling taking place in a stratified medium, we could also consider the coupling taking place in a random medium.

This problem has also been considered by Frisch (1965) in the case of two interacting modes.

The problem is a problem of stochastic equations, which very generally is the problem of wave propagation in random media.

The method is of a very general nature. As an example, we shall consider first an equation of the following form:

$$\frac{\partial \psi}{\partial t} = L_o \psi + L_1 \psi \tag{2.141}$$

where L_1 is a random operator satisfying the condition

$$< L_1 > = 0 \ .$$

ψ has a non-random part ψ_o and a random part $\delta\psi$, $< \delta\psi > = 0$. The random operator, operating on $\delta\psi$, the quantity $< L_1 \delta\psi >$ is different from zero, which means that equation (2.141) is not really a linear equation. We can easily obtain a formal development for ψ_o in the following way.

For $L_1 = 0$, we can write the solution:

$$f(r, t) = e^{L_o t} \ f(r, o) \tag{2.142}$$

This gives the possibility of a formal expansion of ψ:

$$\psi = e^{L_o t}\ \psi(o) + \int_o^t \left(e^{L_o(t-t_1)}\ L_1\ e^{L_o t_1}\ \psi_o\right) dt_1 + \int_o^t dt_1 \int_o^{t_1}$$

$$e^{L_o(t-t_1)}\ L_1\ (t_1)\ e^{L_o(t_1 - t_2)}\ L_1\ (t_2)\ e^{L_o t_2}\ \psi(o)\ dt_2 \qquad (2.143)$$

$+ \ldots\ldots\ldots$

It is possible to take the average of both sides and avoid the difficulties due to similar terms by retaining only the appropriate terms.

The formal solution expresses that the initial function $\psi(r,o)$ has been scattered by the inhomogeneities once, twice, or more often.

After taking the average, by differentiation with respect to t, it is possible to obtain as a first approximation:

$$\frac{\partial < \psi >}{\partial t} = L_o < \psi > + \int_o^t < L_1(t)\ e^{L_o(t-t_1)}\ L_1(t_1) >< \psi\ (t_1) > dt_1 \qquad (2.144)$$

assuming, essentially, that the function ψ does not change appreciably over a correlation time of the operator L_1, and that the terms which have not been written are negligible. This latter condition can be expressed by saying that the function $< \psi >$ does not change appreciably during a correlation time or along a correlation length. In the case of the sound wave equation:

$$\nabla^2 \psi + k_o^2 \left(1 + \mu(r)\right) \psi(r) = 0 \qquad (2.145)$$

where μ is the random part of the index of refraction, it is possible to take as a unit of length the correlation length. Assuming that $\mu = \epsilon\mu^*$, where ϵ is a small quantity, we see that we can write in dimensionless units:

$$\Delta^* \psi(r^*) + k_o^2 L^2 \psi(r^*) + \epsilon k_o^2 L^2 \mu^*(r^*) \psi(r^*) - 0 \qquad (2.146)$$

If the number $\epsilon k_o^2 L^2$ (called by Frisch the generalized Reynolds number) is small, we have a necessary condition for the validity of the first order smoothing approximation.

If we want to consider how the propagation occurs in a random medium, when several modes can propagate, it is useful to consider the eigenfunctions corresponding to the average medium. If the modes of the average medium have eigenfunctions $\psi_j(\vec{k}) \exp(i\,\vec{k}.\vec{\sigma})$, it is possible to develop the solution of the equation of propagation in the form:

$$\psi(\vec{r},t) = \Sigma_j \int d\vec{k}\; \alpha_j(\vec{k},t)\; \Phi_j(\vec{k})\; e^{i\vec{k}.\vec{r}} \qquad (2.147)$$

where $\alpha_j(\vec{k},t)$ is the amplitude of the mode j associated to the wave vector $\vec{k}$.

We shall now use a model of the propagation which can help us to understand the physics of the problem. Let us assume that we can diagonalize the system of equations which the α_j satisfy:

$$\frac{\partial}{\partial t} \alpha_j(\vec{k},t) = i\,\omega_j(\vec{k})\, q_j(\vec{k},t) + \Sigma_{\ell=1}^{n} \int C_{j\ell}(\vec{k},\vec{k}')\, q_j(\vec{k}',t)\, d\vec{k}' \qquad (2.148)$$

where $\omega = \omega_j(\vec{k})$ is the equation of dispersion of the j mode and the $C_{j\ell}$ are the coupling coefficients between the wave modes.

To make things even simpler, we shall consider two mödes only with two equations:

$$\frac{\partial \varphi_1}{\partial t} = i\omega_1(\vec{k})\, \varphi_1(\vec{k}) + \int B_{12}(\vec{k},\vec{k}')\, \varphi_2(\vec{k}')\, d\vec{k}'$$

$$\qquad (2.149)$$

$$\frac{\partial \varphi_2}{\partial t} = i\omega_2(\vec{k})\, \varphi_2(\vec{k}) + \int B_{21}(\vec{k},\vec{k}')\, \varphi_2(\vec{k}')\, d\vec{k}'$$

We have left out terms B_{11} and B_{22} which express self coupling.

The conservation of energy implies:

$$\frac{d}{dt}\int\left[|\varphi_1|^2 + |\varphi_2|^2\right] d\vec{k} = 0 \tag{2.150}$$

This has the consequence that ω_1 and ω_2 are real and that

$$B_{12}\,(\vec{k},\vec{k}') = B^*_{21}\,(\vec{k},\vec{k}') \tag{2.151}$$

The physical process is then the following: the Fourier component k' of the wave 1 interacts with the Fourier component k-k' of the inhomogeneities to produce a component k of the wave 2. The generation of the second mode from the first mode is a multiple diffusion process. We shall first consider the case where the generalized Reynolds number is small. In that case, the mean amplitude of the waves propagate as in a homogeneous medium, with a renormalization of the frequencies:

$$\nu_1\,(\vec{k}) = \omega_1\,(\vec{k}) + i\,\beta_1\,(\vec{k}) \tag{2.152}$$

$$\nu_2\,(\vec{k}) = \omega_2\,(\vec{k}) + i\,\beta_2\,(\vec{k})$$

β_1 and β_2 are complex and imaginary conjugates. β_1 is given by:

$$\beta_1\,(\vec{k}) = i\,\epsilon^2 \lim_{\substack{\eta\to o\\ \eta>o}} \int \frac{r(\vec{k},\vec{k}')\,dk'}{\omega_2(k')-\omega_1(k)+i\eta} \tag{2.153}$$

where ϵ^2 is a characteristic quantity of the relative fluctuations of the parameters $r(\vec{k},\vec{k}')$ in the Fourier transform of the autocorrelation function. Generally $Re(\beta) > o$; the mean amplitudes propagate as if the medium were absorbing. In fact, there is a transfer of energy of mode 1 to mode 2 and, simultaneously a transfer of energy towards the small wavelength. If the generalized Reynolds number is large, we can treat the random parameters as random variables. The coupled equations become now ordinary differential equations with random parameters.

$$\frac{\partial \varphi_1}{\partial t} = i\omega_1 \varphi_1 (t) + i\varepsilon b \varphi_2 (t)$$

$$\frac{\partial \varphi_2}{\partial t} = i\omega_2 \varphi_2 (t) + i\varepsilon b \varphi_1 (t) \tag{2.154}$$

Energy transfer is efficient if $\omega_1 \neq \omega_2$. In that case, with a probability density $\underline{P}(b)$, we obtain for the mean energy of mode 2:

$$E_2 (t) = \frac{1}{2} \int_{-\infty}^{+\infty} (1 - \cos 2bt) \underline{P}(b) \, db \tag{2.155}$$

As t goes to infinity, $E_2 (t) \to \frac{1}{2}$. The asymptotic energy distribution is equipartition.

Let us summarize the results thus obtained. Essentially, if we make a Fourier transform of the random properties of the medium, we can say that the Fourier component k' of mode 1, interacting with the Fourier component k-k' of the inhomogeneities, generates the Fourier component k of mode 2.

Two cases can be considered. The first one corresponds to the case where the wavelength is very large compared to the characteristic scale of the random medium. In such a case, there is a constant feedback between modes 1 and 2, but with a change of spectrum. The interaction with the inhomogeneities feeds the short wavelength (high wave numbers). In the second case, the wavelength is smaller than the characteristic scale of the random medium, and the spectrum remains stationary, but there is a tendency towards equipartition of the energy between the two modes.

This is probably the most important result. If applied to the problem of the chromosphere, this would mean that about one-third of the mechanical energy which reaches the chromospheric level can go into the Alfvén mode. If we take Osterbrock's numbers, this gives the following results (h= 750 km):

Table X

Energy Flux in the Alfvén Mode

($erg/cm^2/sec$)

B (gauss)	$F_+(0)$	$F_+(z)$	$F_{A+}(z)$
0.5	3×10^7	2.6×10^6	0.8×10^6
2	3×10^7	2.5×10^6	0.8×10^6
50	3×10^7	3.4×10^6	1.1×10^6
50	3×10^8	5.6×10^6	1.8×10^6

This leads to the conclusion that there is no difficulty in feeding the Alfvén mode at the chromospheric level, which will provide the amount of mechanical energy necessary to heat the corona.

J. Dynamo Mechanism.

Dissipation of the magnetic field by the Joule effect is a well-known phenomenon. The equation of evolution of a magnetic field

$$\eta \nabla^2 B = \frac{\partial B}{\partial t} \tag{2.156}$$

leads to a time of decay

$$\tau = \frac{L^2}{\eta} \tag{2.157}$$

where L is a characteristic length of the system.

If we had to take for the characteristic size the radius of a star, the time of decay of a magnetic field in a fully ionized plasma, with a resistivity

$$\eta = \frac{\pi^{3/2} m_e^{1/2} Ze^2 c^2 \log \Lambda}{2(2kT)^{3/2}} \qquad (2.158)$$

or

$$\eta = 3.80 \cdot 10^{12} \frac{z \log \Lambda}{T^{3/2}} \quad \text{e.m.n. c.g.s.},$$

is extremely large. In fact, if we take $T = 10^6$, $\eta \simeq 10^3$, $\tau \simeq 10^{18}$ sec $\simeq 3.10^{10}$ years.

In fact, if we have a region of higher resistivity, it determines the lifetime of the magnetic field. The outer layers of a star, like the sun, have a much higher resistivity, of the order of 10^7 to 10^8, which leads to a characteristic time of the order of 10^6 years.

However, this time scale represents the time needed for the inner magnetic field to separate from the outer magnetic field which is evolving toward a vacuum field. This short time scale has to be considered only as far as the external magnetic field is not a vacuum field.

It is such a circumstance which leads to the idea that a dynamo mechanism is necessary to explain the magnetic field of a star like the sun.

In the case of A stars, the surface resistivity is of the order of 10^6, giving a lifetime of the order of 3.10^7 years. The magnetic field of A stars can be a fossil magnetic field, as the lifetime of the magnetic field is comparable to the lifetime of the star. As mentioned by Cowling (1965) further difficulties arise from consideration of convective effects.

In the equation of evolution of the magnetic field,

$$\frac{\partial B}{\partial t} = \vec{\nabla} \times (\vec{v} \times \vec{B} - \frac{1}{\mu_o \sigma} \vec{B}), \qquad (2.159)$$

the ratio of the two effects is of the order of $(4\pi\sigma VL)^{-1}$, where V and L are quantities of the order of the velocity and of the length scale of variation of the field. For $\sigma = 10^{-3}$, $V = 10^2$,

$L = 10^{10}$ cm, this gives a ratio of the order of 10^{-10}, which is very small. We can certainly conclude that the effect of the motions is very important; we shall consider the problem in some detail later.

In earlier work on the dynamo mechanism, it was shown by Cowling (1932) that no axisymmetric velocity field can maintain a magnetic field.

An interesting argument concerning the dynamo mechanism has been given by Bondi and Gold (1950). If lines of force were exactly frozen into the material of a star, no new lines of force could emerge from the surface, and so dynamo action would be unable ever to build up an external field.

The leaking of lines of force, due to finite conductivity, is implicit in many dynamo theories.

The first idea of a turbulent dynamo is due to Parker (1955). It is assumed that, initially, loops of magnetic force are present in meridian planes. The non uniform rotation (supposed to be mainly a variation of angular velocity with depth) then pulls out the vertical parts of the loops in the azimuthal direction, thereby creating an azimuthal field. Next, convective upsurges lift the lines of force of the azimuthal field; because of Coriolis force, the inflow at the base of an upsurge creates a whirling motion round the vertical, so that the convection twists the lines of force about the vertical as well as lifting them (Fig. 16). Hence, new loops of magnetic force are created in meridian planes which supplement the original field in those planes.

The statistical problem of finding the evolution of the average field has made some progress by using the methods of studying random equations. If we consider again the equation of evolution of the magnetic field in the presence of a velocity field,

$$\frac{\partial B}{\partial t} = \nabla \times (\vec{v} \times \vec{B}) + \frac{1}{\mu_o \sigma} \nabla^2 B \tag{2.160}$$

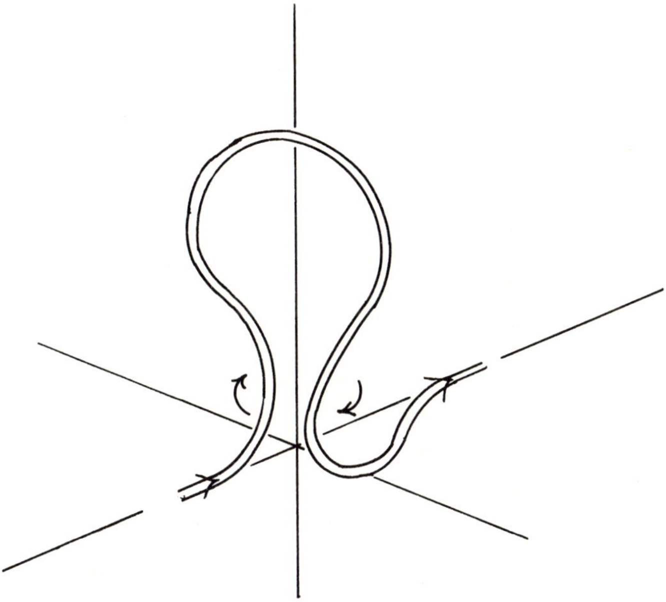

Figure 16

we can separate again the random operator and the non-random operator:

$$\frac{\partial B}{\partial t} = L_o B + L_1 B . \tag{2.161}$$

The average field $< B >$ follows, for the first order smoothing approximation, the following equation:

$$\frac{\partial < B >}{\partial t} = L_o < B > + \int_o^t <L_1(t) e^{L_o(t-t_1)} L_1(t_1)> < B(t_1)> dt_1 \tag{2.162}$$

The only mathematical difficulty is to write explicitly the operator $\exp\ L_o((t-t_1))$. When the average velocity field vanishes, and for an infinite medium this is not a difficulty, as the Green functions of the non-random operator can be easily written,

$$G(\vec{x},t) = \left(\frac{\mu\sigma}{4\pi t}\right)^{3/2} \exp\left(-\frac{\mu\sigma}{4t} x_\ell x_\ell\right). \tag{2.163}$$

Problems are more complicated when the average velocity field is non-zero and when there are boundaries. The problem, however, can be still solved when the system is not far from being homogeneous and isotropic, and when the boundaries can be considered as very far compared to the scale of the turbulence. The average of the product of the random operator reflects directly the properties of the auto-correlation function of the velocity field.

Batchelor (1952), following the method of Robertson (1940), shows that the auto-correlation tensor of the velocity field takes for the homogeneous and isotropic turbulence the form:

$$R_{ik} = < V_i(x,t) V_k(x',t') > = A(\xi)\delta_{ik} + B(\xi)\ \xi_i \xi_k + C(\xi)\ \epsilon_{ik\ell}\xi_\ell \tag{2.164}$$

where $A(\xi)$, $B(\xi)$, $C(\xi)$ are functions only of the distance ξ

between the two points x and x'. In the case of spherical symmetry, C = 0; if there is no mirror symmetry, C ≠ 0. This corresponds to the case where there is a term which changes sign when the directions of one of the axes of coordinates is reversed. As an example of turbulence with no mirror symmetry, we can consider turbulence in the presence of rotation.

It is found, in the case of isotropic turbulence with no mirror symmetry, that the integral term can be written:

$$S = \nabla^2 (\nu_T \vec{B}) + \vec{\nabla} \times (\alpha \vec{B}) \tag{2.165}$$

with the following values of the terms ν_T and α:

$$\begin{aligned} \nu_T &= \frac{4\pi}{3} \int_0^t \int_0^\infty (3B + A\xi^2)\xi^2 \, G d\xi d\tau \\ \alpha &= -\frac{8\pi}{3} \int_0^t \int_0^\infty \left(\xi \frac{\partial C}{\partial \xi} + 3C\right) \xi^2 G d\xi d\tau \end{aligned} \tag{2.166}$$

The following choice can be made for A, B, and C:

$$\left\{ \begin{aligned} A(\xi,\tau) &= \epsilon \frac{\lambda^2}{\tau_o^2} f\left(\frac{\xi}{\lambda}\right) g\left(\frac{\tau}{\tau_o}\right) \\ B(\xi,\tau) &= \beta \frac{1}{\tau_o^2} f\left(\frac{\xi}{\lambda}\right) g\left(\frac{\tau}{\tau_o}\right) \\ C(\xi,\tau) &= \gamma \frac{\lambda}{\tau_o^2} f\left(\frac{\xi}{\lambda}\right) g\left(\frac{\tau}{\tau_o}\right) \end{aligned} \right. \tag{2.167}$$

where ϵ, β, γ are dimensionless quantities. The functions f and g express the properties of the space correlations and time correlations

A convenient but arbitrary choice is the following:

$$\begin{cases} f = \exp\left(-\dfrac{\xi^2}{\lambda^2}\right) \\ g = \exp\left(-\dfrac{\tau}{\tau_o}\right) \end{cases} \tag{2.168}$$

Considering the value of the conductivity σ, the following inequality is satisfied:

$$\frac{\mu\lambda^2\sigma}{4\tau_o} \gg 1 \quad .$$

This leads to the following values of α and ν_T :

$$\begin{aligned} \alpha &= -2\gamma \frac{\lambda}{\tau_o} \\ \nu_T &= \epsilon \frac{\lambda^2}{\tau_o} + 2\beta \ \nu_R \end{aligned} \tag{2.169}$$

The condition

$$\frac{\lambda_o^2 \ \mu\sigma}{4\tau_o} \gg 1$$

expresses the fact that the correlation time of the turbulence is much smaller than the diffusion time of the magnetic field over a correlation length.

Only an estimation can be made of the coefficients ϵ, β, γ. Knowing the mean square value u^2 of the velocity, Batchelor gives, as an order of magnitude:

$$\beta \cong \frac{u^2 \tau_o^2}{2\lambda^2}$$
$$\epsilon \cong \frac{u\,\tau_o}{\lambda^2} \tag{2.170}$$

Finally, we can obtain an estimate of γ in the following way. Differentiating R_{ik} with respect to ξ_m, we obtain:

$$R_{ik,m} = < v'_i(x,t) . v'_{k,m}(x + \vec{\xi},t) >$$
$$= \frac{1}{\xi}\frac{dA}{d\xi}\delta_{ik}\xi_m + \frac{1}{\xi}\frac{dB}{d\xi}\xi_i\xi_k\xi_m + B(\delta_{im}\xi_k + \delta_{km}\xi_i)$$
$$+ \frac{1}{\xi}\frac{dc}{d\xi}\epsilon_{ik\ell}\,\xi_\ell\,\xi_m + C\,\epsilon_{ikm}. \tag{2.171}$$

Multiplying both sides by ϵ_{ikm}, we obtain:

$$2\xi\frac{dC}{d\xi} + 6C = - < w'(\vec{x},t)\ \nabla x w'(\vec{x} + \vec{\xi},t) > \tag{2.172}$$

and especially

$$C(o) = -\frac{1}{6} < w'(\vec{x},t)\ .\ \nabla\ x\ w'(\vec{x},t) > \tag{2.173}$$

from which we derive γ:

$$\gamma = \frac{1}{6}\ \frac{< w(x,t).\ \ \nabla\ x\ w(x,t)\ \ > \tau_o^2}{\lambda} \tag{2.174}$$

To get an order of magnitude, we shall also suppose that $< w.\nabla\ x\ w >$ is of the same order of magnitude as $2\ \bar{w}\ \omega$, where $\bar{w}$ is the same average value of the velocity along the vertical. In the case of the solar convective zone, we get the following estimates:

Table XI

Turbulence Parameters

λ (meters)	τ_o(sec)	σ(mho/m)	ϵ	β	γ	$\frac{\lambda^2}{4\nu\tau_o}$
$1.8\cdot10^6$	100	10^2	$2\cdot10^{-4}$	10^{-4}	$2\cdot10^{-6}$	$1.2\cdot10^6$
$3\cdot10^6$	400	10^3	$1.6\cdot10^{-2}$	$8\cdot10^{-3}$	$5\cdot10^{-5}$	$7\cdot10^6$
$3\cdot10^7$	50000	10^{-2}	0.7	0.35	$8\cdot10^{-2}$	

U(m $\sec^{-1}$)	α
$3\cdot10^2$	0.03
10^3	0.4
$5\cdot10^2$	25.

The turbulent diffusivity $\nu_T \simeq 10^9$ to $10^{11}\,cm^2 sec^{-1}$ is 10^5 to 10^7 larger than the microscopic diffusivity.

The equation of the turbulent dynamo takes then the form:

$$\frac{d\vec{B}}{dt} = (\nu + \nu_T)\ \nabla^2\vec{B} + \alpha\ \vec{\nabla} \times \vec{B} \tag{2.175}$$

For an infinite medium, a plane parallel wave is the solution of the equation, with an equation of dispersion:

$$i\omega = -(\nu + \nu_T)\ k^2 \pm \alpha\ k \tag{2.176}$$

For any k smaller than the scale λ, the rate of growth of the magnetic field is of the order of αk. If we take, as an order of magnitude, $k \simeq 10^{-10}$, we obtain $|i\omega| \simeq 2.5\ 10^{-9}$. This corresponds to a time of growth of the order of 15 years for the whole outer convective zone and a dimension of the order of the sun.

More generally, if G is the Green function of the non-random part of the equation, with

$$\alpha = -2 \int_0^t \int_{-\infty}^{+\infty} G\left(\rho_k \frac{\partial C}{\partial \rho_k} + C\right) d\vec{\rho}\, d\tau$$

$$\nu_T = \int_0^t \int_{-\infty}^{+\infty} G(A + B\,\rho_k^2)\, d\vec{\rho}\, d\tau \qquad (2.177)$$

the equation for the average field is:

$$\frac{\partial}{\partial t} < \vec{B} > = \nu\, \nabla^2 < \vec{B} > + \vec{\nabla} \times (< \vec{v} > \times < \vec{B} >) + \alpha\, \vec{\nabla} \times < \vec{B} > \qquad (2.178)$$

This is of the same form as the equation derived by Krause et al (1966), but the coefficient α is derived here in a more rigorous way.

As examples of a solution of the dynamo mechanism, we can consider the following cases:

(1) The force free field.

A force free field,

$$\vec{\nabla} \times \vec{B} = \beta\, \vec{B} \qquad (2.179)$$

is a solution. With $\frac{\partial \vec{B}}{\partial t} = \lambda\, \vec{B}$, we have

$$\lambda = -(\nu + \nu_T)\, \beta^2 + \alpha\beta \qquad (2.180)$$

We can see that a force free field can be stationary if

$$\beta = \frac{\alpha}{\nu + \nu_T} < 0 \qquad (2.181)$$

The field increases only if the current and the field have opposite directions, and if the scale of the field is sufficiently large.

(2) The turbulent slab.

This has some similarity to the convective zone of a star. We consider the two equations:

$$\frac{\partial B}{\partial t} = \nu \, \Delta \, B \times \alpha \, \nabla \times B \qquad \text{inside the slab}$$

$$\frac{\partial B}{\partial t} = \nu' \, B \qquad \text{outside the slab} \tag{2.182}$$

The problem is determined by the following boundary conditions:

-----The magnetic field vanishes at infinity.

-----The normal component of B is continuous at the interface of the slab.

-----The tangential components of the electric field are continuous at the interface of the slab.

The equation of dispersion is fairly complicated. It is interesting to study the shape of the magnetic field in the stationary case (Fig.17). It remains to justify the approximation which has been used. It is acceptable if the increase of the field during a correlation time is small compared to the field. In other words, the quantity $\frac{\tau_o u}{\lambda}$ has to be small. This quantity is of the order of 10^{-2}. The assumption that the equation of the turbulent dynamo is valid is acceptable.

K. Stopping Convection.

The structure of the magnetic field has some similarity to a force free magnetic field. The pitch angle of the lines of force changes with the distance to the surface.

We can get an idea of the stabilization properties by

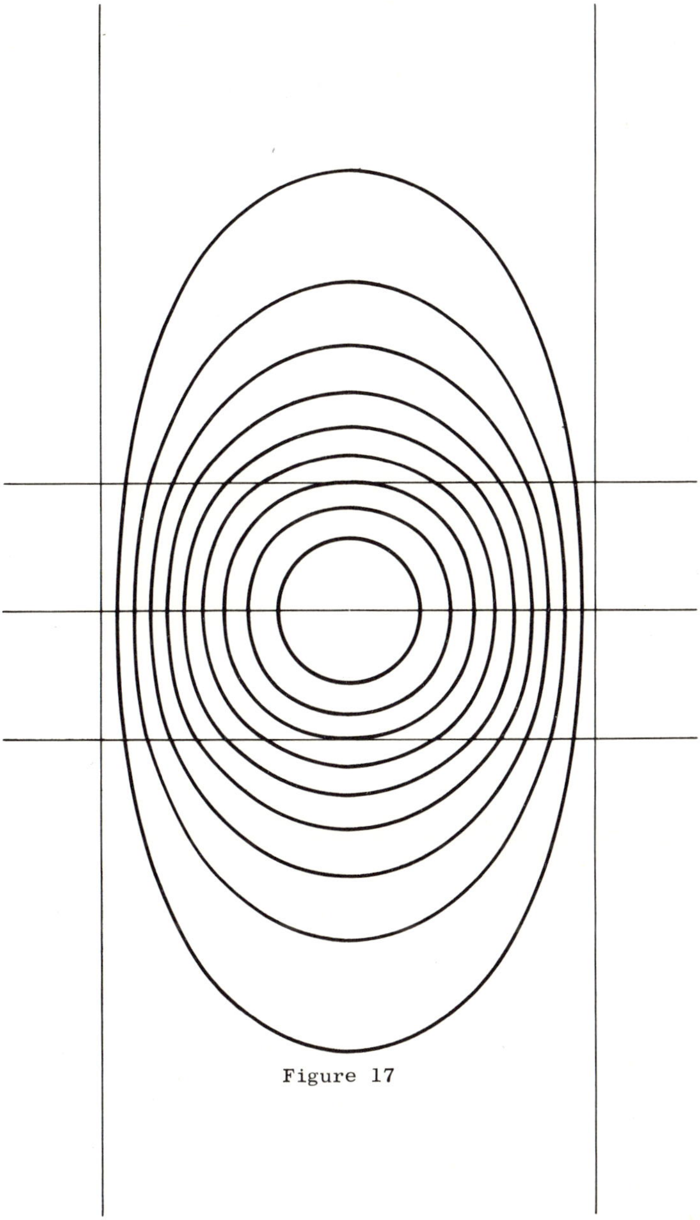

Figure 17

starting from the existing work on stability.

It is well known that a magnetic field retards convection.

Danielson (1961) has used the linearized equation derived by Chandrasekhar (1952, 1960), to study the convection by thermal instability in a conducting fluid in the presence of a magnetic field.

The layer of fluid has a depth d. Davidson has studied the stability of the layer as a function of the magnetic field.

Two numbers have an important physical meaning:

i. The Rayleigh number, defined by:

$$R = \frac{|\beta| \; \gamma d^4}{k \; \nu} \tag{2.183}$$

where $|\beta|$ is the temperature gradient $|\frac{dT}{dz}|$, k the radiative conductivity, ν the kinematic viscosity, and

$$\gamma = g\alpha \tag{2.184}$$

is the product of the acceleration due to gravity and the volume coefficient of expansion.

If we take the equation of motion,

$$\rho \frac{d\vec{v}}{dt} = - \vec{\nabla} p - \vec{g} \rho - \rho \nu \nabla^2 \vec{\mathbf{v}} \; ,$$

the Rayleigh number can be understood as the ratio of the viscous force to the buoyancy force. The viscous force is of the order of

$$\rho \nu \frac{v}{d^2}$$

where d is a characteristic length. The velocity is of the order of d divided by a time t. It comes out that the characteristic time is the time needed for the heat transport over a distance d:

$$t = \frac{d^2}{k}$$

On the other hand, with a dilatation coefficient α, such that

$$\frac{\delta\rho}{\rho} = \alpha\delta T$$

the buoyancy force is $g\rho\alpha\delta T$.

The temperature difference δT is proportional to the temperature gradient $\beta = \left|\frac{dT}{dz}\right|$:

$$\delta T = \beta d.$$

We then have for the viscous force

$$f_{vis} = \rho\nu k d^{-3}$$

and for the buoyancy force

$$f_{buoy} = g\rho\alpha\beta d$$

Convective motion takes place when $f_{buoy} \gg f_{vis}$, which gives

$$\frac{g\alpha\beta d^4}{k\nu} \gg 1$$

For free boundaries, the critical value is of the order of 1000. Experiments with solid boundaries (quoted in Chandrasekhar [1961]) are quite in agreement with the theory.

ii. The magnetic number Q.

$$Q = \frac{\mu d^2}{4\pi\rho\eta\nu} H_{eff}^2 \tag{2.185}$$

where μ is the magnetic permeability and η the electrical resistivity. H_{eff} is related to the actual field in the following way:

The magnetic Reynolds number is related to the ratio of the viscous force to the magnetic force; but this time the characteristic time is the diffusion time for the lines of force,

$$t = d^2\eta^{-1}$$

where η is the resistivity. The magnetic force is larger than the viscous force when

$$\frac{H^2}{d} \gg \frac{\rho\eta\nu}{d^3}$$

which introduces the magnetic Reynolds number.

(a) For a vertical field:

$$H_{eff} = H \tag{2.186}$$

(b) For a horizontal field and convection cells of length ℓ:

$$H_{eff} = H \frac{d}{\ell} \tag{2.187}$$

However, a horizontal field cannot stop a transverse convection, as it cannot stop the exchange instability.

(c) For an arbitrary direction of the magnetic field, but for infinitely long convection rolls,

$$H_{eff} = H \cos \theta \tag{2.188}$$

where θ is the angle of the field with the vertical.

Another parameter which defines the condition of stability is a geometrical parameter defining the cell shape:

$$y = \frac{k_x^2 + k_y^2}{k_z^2} + 1 \tag{2.189}$$

with $k_z = \frac{\pi}{d}$.

We are essentially interested in the qualitative aspect of the problem, and it will be sufficient to consider the curves separating the plane $\log_{10}Q$, $\log_{10}k$ for a special value of the geometrical parameter y.

Figure 18 has been obtained for the following values of the physical constants, corresponding more or less to the physical conditions of the solar photosphere:

$$d = 3 \times 10^7 \text{ cm} = 300 \text{ km}$$
$$K = 10^{12} \text{ cm}^2 \text{ sec}^{-1}$$
$$\eta = 4 \times 10^6 \text{ cm}^2 \text{ sec}^{-1}$$
$$\nu = 10^3 \text{ cm}^2 \text{ sec}^{-1}$$

The values of R and Q (for H = 1 gauss) are log R = 11.3; log Q = 10.3.

It is quite clear than an increase in Q and a decrease in R would place one in the region of stability (Fig. 18).

If we consider a constant magnetic field and a constant superadiabatic gradient:

$$\beta = \left| \left(\frac{dT}{dz}\right)_{ad} - \left(\frac{dT}{dz}\right)_{actual} \right| \tag{2.190}$$

we can follow the point (Q,R) through the atmosphere, assuming for example, a constant value of the characteristic height d.

The Rayleigh number changes very much with height, increasing roughly with ρ, whereas the magnetic number changes very little, the product $\rho\nu$ being almost constant.

When going down in the atmosphere, we are likely to successively find three different regions:

------the stable region, where the magnetic field completely stabilizes the atmosphere;

------the overstable region, where growing oscillation can take place;

------the unstable region, where the energy density of matter is such that convection can actually take place.

The preceding treatment is admittedly very rough, but nevertheless suggests that in a star with strong magnetic field, the convection is still dominating in the lower layers, but is completely inhibited in the upper layers.

In the presence of a magnetic field with a varying pitch angle, the stabilization can be due to the change of pitch with height. From elementary arguments on the tension of the lines of force, it is possible to obtain a condition of stability. We shall compare the buoyancy force to the magnetic force which results from the bending of the lines of force.

The change in azimuthal direction of the lines of force, as a function of height results in a current proportional to

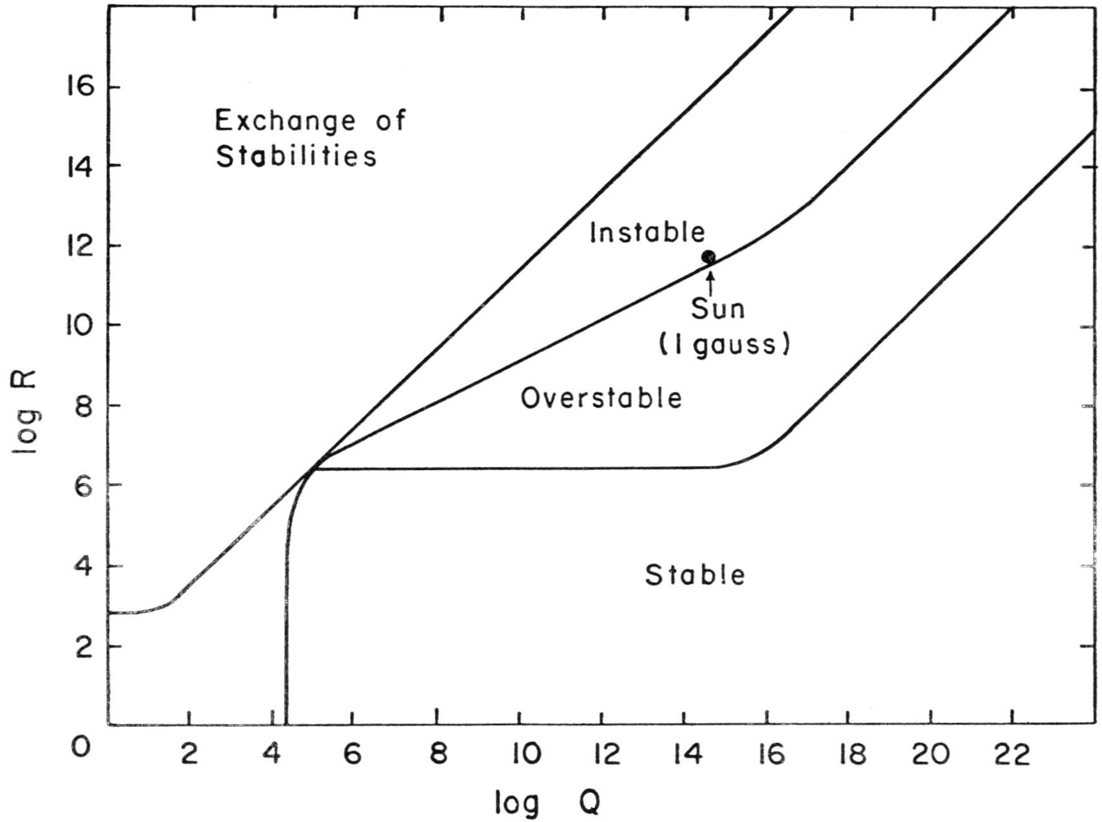
log R
16
14
12
10
8
6
4
2
0
2
4
6
8
10
12
14
16
18
20
22
log Q
Exchange of Stabilities
Instable
Sun
(1 gauss)
Overstable
Stable

Figure 18

$\frac{\partial B_x}{\partial z}$. With an horizontal wave number k, the perturbation of B_x is of the order of $k \xi B$, where ξ is the displacement. The restoring force is then of the order of $k \xi B_x \frac{\partial B_x}{\partial z}$. On the other hand, the buoyancy force will be of the order of:

$$\frac{\xi g \rho}{H} \frac{\gamma - 1}{\gamma} \left(1 - \frac{\nabla^*}{\nabla_{ad}}\right)$$

. With $g\rho = -\frac{dP}{dz} = \frac{-P}{H}$ where H is the pressure scale height, we obtain the stability condition

$$\frac{1}{\mu_o} k B_x \frac{\partial B_x}{\partial z} > \frac{P}{H^2} \frac{\gamma - 1}{\gamma} \left(\frac{\nabla^*}{\nabla_{ad}} - 1\right) \qquad (2.191)$$

BIBLIOGRAPHY

Chapter II

ATHAY, R. G., (1967), "Radiation Energy Loss from the Solar Chromosphere and Corona," Ap. J., 146, 223.

BAKER, N., (1963), "Tables of Convective Stellar Envelope Models," Goddard Space Flight Center.

BATCHELOR, G. K., (1953), "The Theory of Homogeneous Turbulence," Cambridge University Press.

BILLINGS, D. E., (1966), "A Guide to the Solar Corona," Academic Press.

BÖHM, K. H., (1966), "Die Lithium-Häufigkeit und die Structur der äusseren Konvektionszone in sonnenähnlichen Sternen," Z. F. Naturforschung, 21 a, 1107.

BONDI, H., and GOLD,T. (1950), "On the Generation of Magnetism by Fluid Motion," M. N., 110, 607.

BURBIDGE, G. and BURBIDGE,M., (1955), "An Analysis of the Magnetic Variable a canum Venaticorum,"Ap. J., Suppl., 1, 431.

CHANDRASEKHAR, S., (1961), "Hydrodynamic and Hydromagnetic Stability," Oxford University Press.

COWLING, T. G., (1965), Magnetic Stars in "Stellar Structure," Vol VIII of "Stars and Stellar Systems," p. 425, The University of Chicago Press.

DANIELSON, R. E., (1961),"The Structure of Sunspot Penumbras II." Theoretical. Ap. J., 134, 289.

EDDINGTON, A. S., (1926), "The Internal Constitution of the Stars," Chap. XII, Cambridge University Press.

FESSENKOFF, V. G., (1949), "On the Problem of Cosmogony in Contemporary Astronomy," A. J. U.S.S.R. 27, 67.

FRISCH, U., (1964), "Coupling of Magnetodynamic Waves in a Stratified Media," Ann. d'Ap., 27, 224.

FRISCH, U., (1965), "Propagation et couplage d'ondes dans un plasma turbulent," Jour. Phys. Rad., 26, 507.

FRISCH, U. (1966), "La propagation des ondes en milieu aléatoire et les équations stochastiques,"1ene partic, Ann. d'Ap., 29, 645.

FRISCH, U., (1967), "La propagation des ondes en milieu aléatoire et les équations stochastiques, II," Applications, Ann. d'Ap., 30, 565.

KAHALAS, S. L., and McNEILL, D. A., (1964), "Coupling of Magnetohydrodynamic to Electro-magnetic and Acoustic Waves at a Plasma-Neutral Gas Interface," Phys. of Fluids, 7, 1321.

KATO, S., (1966), "On the Atmospheric Oscillations Excited by Turbulence," Ap. J., 143, 372.

KOHL, K., (1966), "Das Eindringen von Strömungen in die stabile Schicht unterhalb der solaren Wasserstoffkonvektionszone," Z. f. Ap., 64, 472.

KRAUSE, F., RÄDLER, K. H., and SEENBECK, M., (1966), "Berechnung der mittleren Lorentz-Feldstärke v x B für ein elektrisch leitendes Medium in turbulenter, durch Coriolis-Kräfte beeinflusster Bewegung," Zs. f. Naturf., 21 a, 369.

KUHI, L. V., (1964), "Mass Loss from T Tauri Stars," Ap. J., 140, 1409.

KUPERUS, M., and ATHAY, R. G., (1967), "On the Origin of Spicules in the Chromosphere - Corona Transition Region," Solar Physics 1 (3/4), 341.

LEIGHTON, R. B., (1959), "Observations of the Solar Magnetic Field in Plage Regions, Ap. J., 130, 360.

LEIGHTON, R. B., (1964), "Transport of Magnetic Field on the Sun," Ap. J., 140, 1547.

LIGHTHILL, M. J., (1952), "On Sound Generated Aerodynamically," Proc. Roy. Soc. A., 211, 564.

MASSEVITCH, A. G., (1949), "Stellar Evolution Accompanied by Corpuscular Radiation from the Standpoint of the Theory of Internal Stellar Structure," A. J. U.S.S.R., 26, 207.

MEIN, P., (1966), "Champ macroscopique des vitesses dans l'atmosphère Solaire d'après les mesures de déplacement des raies de Fraunhofer," Ann. d'Ap., 26, 153.

OSTERBROCK, D. E., (1961), "The Heating of the Solar Chromosphere, Plages, and Corona by Magnetohydrodynamic Waves," Ap. J., 134, 347.

PARKER, E. N., (1955), "Hydromagnetic Dynamo Models," Ap. J., 122, 293.

PARKER, E. N., (1963), "Kinematical Hydromagnetic Theory and its Application to the Low Solar Photosphere," Ap. J., 138, 552.

PIDDINGTON, J. N., (1956), "Solar Atmospheric Heating by Hydromagnetic Waves," M.N.R.A.S., 116, 314.

PIKELNER, S. B., (1962), "Formation of Chromospheric Network and the Structure of the Magnetic Field," A. J. U.S.S.R., 6, 757.

PROUDMAN, I. (1952), "The Generation of Noise by Isotropic Turbulence," Proc. Roy. Soc., A 214, 119.

ROBERTSON, H. P., (1940), "The Invariant Theory of Isotropic Turbulence," Proc. Camb. Phil. Soc., 36, 209.

SACK, N., (1966), "On the Fitting of Radiative Layers to Convective Zones in Stellar Atmospheres," Ann. d'Ap., 29, 633

SCHATZMAN, E., (1964), "The Heating of the Solar Corona and Chromosphere, Ann d'Ap., 12, 203.

SCHATZMAN, E., (1954a), "Remarques sur l'évolution des étoiles, III," Ann d'Ap., 17, 300.

SCHATZMAN, E., (1954b), "Remarques sur l'évolution des étoiles, IV," Ann. d'Ap. 17, 377.

SCHATZMAN, E., (1954c), "Remarques sur l'évolution des étoiles, V," Ann. d'Ap. 17, 382.

SCHATZMAN, E., (1964), "On a Simple Hydrodynamic Problem with Astrophysical Applications," Astrophysica Norvegica, 9, 283.

SCHATZMAN, E., and SOUFFRIN, P.,(1967), "Waves in the Solar Atmosphere," Ann. Rev. Ast. and Ap., 5, 67.

SCHWARZSCHILD, K., (1906),"Über das Gleichgewicht der Sonnenatmosphäre," Göttinger Nachr., 41.

SCHWARZSCHILD, M., and RICHARDSON, R. S., (1950), "On the Turbulent Velocities of Solar Granules," Ap. J., 111, 351.

SIMON, G. W., and LEIGHTON, R. B., "Velocity Fields in the Solar Atmosphere, III, Large Scale Motions, the Chromospheric Network and Magnetic Fields," Ap. J., 140, 1120.

SEVERNY, A., (1965a), "Solar Magnetic Fields," Trans. I.A.U., XII, B, 755.

SEVERNY, A., (1965b), "Sunspots: Observations," I.A.U. Symposium, 22, 238.

SKALAFURIS, A. (1968), Ap. Letters.

SOUFFRIN, P., (1966), "Hydrodynamique d'une atmosphère perturbée par une zone convective turbulente sous-jacente," Ann. d'Ap., 29, 55.

SPITZER, L. JR., (1959), "Influence of Fluid Motions on the Decay of an External Magnetic Field, Ap. J., 125, 525.

UCHIDA, Y., (1967), "Resonant Responses of the Solar Atmosphere to the Gravitational-Hydrodynamic Waves," Ap. J., 147, 181.

UNNO, W., (1964), "Generation of Acoustic Noise in Convective Zones," Trans. I.A.U. XII B, 555.

WEYMANN, R., (1960), Heating of Solar Chromosphere by Shock Waves, <u>Ap. J.</u>, 132, 452.

WEISS, N. O., (1964), "Convection in the Presence of Constraints," <u>Phil. Trans. Roy. Soc.,</u> A 256, 99.

III. Solar and Stellar Activity

A. Flare Stars.

According to Haro (1960), observation of T Tauri stars and flare stars shows that they form a continuous sequence.

A typical flare star is UV Ceti. Extensively studied by Oskanjan (1953), UV Ceti is a dM5e star, which shows extremely large light variations. The time involved in such outbursts is of the order of a few minutes; the light increase is usually of 3 magnitudes, but can reach 6 magnitudes. Joy and Humason (1949) in 1948 on Mount Wilson obtained a spectrogram of UV Ceti at, fortuitously, the exact time of an outburst. The normal dM5e spectrum was completely dominated by a spectrum showing (Joy, 1957):

(a) a strong continuum corresponding to a temperature higher than 10,000°K;
(b) strong hydrogen emission lines;
(c) bright lines of He I and a faint λ 4686 line of He II.

Herbig (1957), (1960) reviewed the criteria of T Tauri stars. The T Tauri stars have three general properties:

(a) the possession of a distinctive, bright line spectrum,
(b) the association with nebulosity,
(c) the presence of irregular fluctuations in light.

The irregular fluctuations in light can be either of relatively small amplitude (one magnitude) and of short duration, or of large amplitude and comparatively slow duration. Many of those stars show strengthened hydrogen and continuous emission in the 3600 Å region, which probably accompany the flare-like activity.

Schatzman (1958), suggested that the flare activity should be accompanied by an emission of radio waves. At the symposium on radio-astronomy in July, 1958, Schatzman (1959) gave a quantitative estimate of the flux which should be expected from flare stars, assuming the same efficiency in radio-wave production as on the sun. For the first time, Lovell (1964) measured the radio emission from the flare star UV Ceti in September, 1958. He showed that flare stars have a much greater efficiency in the production of radio waves than the sun, the ratio of the total energy emitted in the radio-range to total energy emitted in the optical range being much higher in flare stars.

$$\left(\frac{W_{radio}}{W_{optical}}\right)_* \simeq 10^{-2} \text{ to } 10^{-3}, \quad \left(\frac{W_{radio}}{W_{optical}}\right)_\odot \simeq 10^{-5}$$

Flare stars show type I, II and III bursts. These results suggest very strongly that flare stars show in a giant form the same kind of activity as the sun. We shall assume in the following that T Tauri stars also present a surface electromagnetic activity. This assumption is substantiated by the fact that the T Tau stars studied by Bonsack and Greenstein (1960) show a high abundance of lithium. It seems likely that lithium was formed by spallation of C^{12}, N^{14}, O^{16}, at the surface of these stars by relatively low-energy cosmic ray-protons. This possibility was considered in some detail by Bernas, Gradsztajn, Reeves and Schatzman (1967).

It seems, therefore, that the understanding of the processes taking place at the surface of active stars will be clear only if a clear understanding of the solar activity is obtained. On the other hand, one should not underestimate the possibility that with a difference in physical conditions, the information concerning flare stars could help in understanding solar activity.

B. Solar Flares, Optical Problems.

The problems of solar flares have been reviewed several times, and the author does not intend to repeat such a review. Solar flares appear first as two bright spots, one on each side of the line of zero longitudinal field. The flare extends later to larger areas. However, the area of the optical flare remains smaller than the total area of the plage in which it takes place.

The total area of the optical flare is about one or two orders of magnitude smaller than the total area of the disturbed region. For example, from the photographs published by Howard and Babcock (1960), we can estimate the area of the disturbed region around the 3^+ flare of July 16, 1959, to be 10^{21} cm^2, whereas the area of the optical flare at its maximum did not exceed 6×10^{19} cm^2. Let us, in particular, consider the flash phase of a solar flare. The brightness of the flaring region increases slowly; then, it suddenly shows a very rapid increase. Based on the precision of observations, the beginning of the flash phase coincided with the appearance of a type III burst. We can consider that the time scale appearance of the flare is of the order of 1 to 10^2 seconds. The interpretation of the flare spectrum suggests that, in the flaring region, densities are of the order of 10^{13} particles cm^{-3}, and temperatures are 1.5×10^4 °K. The densities are much higher, and the temperature much lower than in the surrounding corona at the same level. It was further shown by Suemoto and Hiei (1959), Hirayama (1961), Svetska (1962) (1965), and de Feiter (1964), that Balmer line emission originates in very small structures of 1 to 10 kilometers in diameter. It has often been suggested that the matter which becomes visible in the optical flare is collected there by a mass motion. However, in the following we shall consider the possibility that matter pre-exists prior to the optical flare and becomes visible at the time of the flare by a change in heat balance. To this effect, let us consider the interpretation of filamentary structure in solar prominences by Doherty and Menzel (1965). They consider isobaric steady-state models of prominence as filamentary structures similar to the models of Orrall and Zirker (1961). The models have cylindrical symmetry and contain uniform longitudinal magnetic fields. Thermal conduction transverse to the field balances the net loss of radiative energy at each point of the filament. They have examples of filament structure with various values of the product $N_T T$, in which N_T is the total particle density. With $N_T T = 2.10^{15}$ cm^{-3}, and a magnetic field B = 0.01 gauss, they have a model of a filament of 6 kilometers in diameter and a central temperature of about 10^4 °K. It is undoubtedly possible to build models of filamentary structures with higher central densities. We shall generalize the picture of Doherty and Menzel, by assuming that matter pre-exists in the outer layers of the sun in a filamentary structure and that the filaments are hanging in the magnetic field, the filamentary structure being essentially parallel to the lines of force of the magnetic field.

The possibility that a filamentary structure could be hanging in a magnetic field was considered by Dungey (1958). The lines of force are transverse to the filament, but circle around the filament. The heat transfer problem is the same as in the model of pre-existing filamentary structure, as it seems more likely that the magnetic field is parallel to the filamentary structure. Photographs such as those of Beckers (1963) clearly show dark and bright filamentary structures of 1400 km width and 15,000 to 30,000 km length. It is quite possible to imagine that these filamentary structures are actually made of an assembly of filaments of a smaller diameter. These structures are likely to be spicules projected on the solar disc. In active regions the structure is much more confused. In the solar chromosphere, however, it seems that we can find structures aligned by the magnetic field, which are of a higher density and lower temperature than the surrounding corona.

As yet, no theory exists on the heat balance of these filamentary structures. We shall assume here that the heat balance is essentially determined by the strength of the magnetic field. The magnetic field can determine the heat flow from the surrounding corona. The magnetic field can also be important for determining the rate of transformation of mechanical energy into heat inside the filamentary structure.

C. Mechanism of the Flash Phase.

We shall assume in the following that the triggering mechanism of the solar flares exists; the triggering mechanism itself will be discussed later. Here we are concerned with the interpretation of the flash phase.

We shall assume that the flash phase begins when a decompression wave reaches the region of the filamentary structure in the chromosphere.

If the heat balance of the filaments is essentially determined by the heat flow from the corona, a change of the intensity of the magnetic field will produce a change in the heat flow. The time needed for the adjustment of the filament to the new conditions of heat flow is of the order of the characteristic time for diffusion of heat.

If K is the coefficient of heat conductivity, N, the particle density, and L, a characteristic length, the characteristic time of heat transfer, as estimated by Schatzman (1966), is

$$t = \frac{L^2 Nk}{K} \tag{3.13}$$

Let us use the numbers given by Doherty and Menzel (1965), with $L \simeq 10^5$ cm, $N \simeq 10^{10}$ cm^{-3}, $K \simeq 10^4$ to 10^2 erg cm^{-1}deg^{-1}sec^{-1}, we find characteristic times of the order of 1 to 100 seconds. Even using the densities and temperatures given by Athay and Thomas (1961) for the cold component of the chromosphere, i.e., $N \simeq 6 \times 10^{11} - 10^{12}$ cm^{-3}, K (for a weak magnetic field) $\simeq 10^4$, erg cm^{-1}deg^{-1}-sec^{-1}, but using more strictly the scale depth given by Doherty and Menzel, $L \simeq 4 \times 10^4$ cm, the characteristic time for the adjustment of the heat flow also turns out to be of the order of 1 to 100 seconds. The heat flow in the region of strong temperature gradient can be of the order of 10^4 to 10^6 ergs cm^{-2}. For a filament of 10^6 cm thickness, this corresponds to a total energy available in 10^3 seconds of the order of 10 to 1,000 ergs cm^{-3} which, in this assumption, is driven from the corona to the filamentary structure, and cannot be obtained from the internal energy of the corona, which is much too small. At the level of the chromosphere flares, the internal energy is of the order of one erg cm^{-3}. It is quite clear that the energy has to be supplied continuously from the corona to the filamentary structure.

Whatever the nature of the energy supply of the solar flares is, the energy has to be converted into excitation inside the filamentary structure, and this conversion has to take place along the duration of the entire flare. We must explain not only the very quick start of the flash phase, but also the duration of the flare itself. This is the reason it is suggested here that the heat flow from the corona to the filamentary structure plays an essential role.

Let us consider the energy supply in a disturbed region during the duration of a flare. For example, if we consider again the 3^+ flare of July 16, 1959, the area of the disturbed region is of the order of 10^{21} cm^2. The flux of mechanical energy in the corona is of the order of 10^6 to 10^7 ergs cm^{-2} sec^{-1}. The total duration of the flare was 10^4 seconds. The total energy input in the corona above the disturbed region, during the flare, was 10^{31} to 10^{32} ergs. It is quite clear that a small change in the heat balance in the disturbed area can produce the optical flare.

In the region of low magnetic field, the heat conductivity in the corona is about 5×10^9 erg $cm^{-1}deg^{-1}sec^{-1}$. With a number density of 10^8 cm^{-3}, the heat flow from the corona to the filamentary structures adjust in a time of the order of 300 seconds for $L \simeq 3 \times 10^5$ km. It is therefore quite possible to adjust the heating of the flaring region to the heat flow from the corona.

The motion of the flowing region will be associated with the change in intensity and in shape of the magnetic field. The total duration of the flare can be associated with the time needed for restoring the original magnetic field.

D. Instability Problems.

A crucial problem of stellar flares and solar flares, is the nature of the instability. There is no doubt that the starting of a flare is due to some kind of instability.

A number of instabilities have been considered in the literature. Essentially, one group of instabilities is related to the neutral point instability; the other one is related to resistive instability.

As far as neutral point instability is concerned, we can consider a series of improvements, starting from the first idea developed by Giovanelli (1947).

Dungey (1958) shows, in the case of a pressure free gas with an infinite conductivity that a neutral point is unstable, and that the Lenz law is violated.

The effect of the finite resistivity can be understood by considering the motion of the lines of force with respect to the matter. The drift velocity is

$$\vec{W} = \eta \frac{\vec{j} \times \vec{B}}{B^2} \tag{3.1}$$

and has a singularity at the neutral point. The drift velocity is large only where the current is large. The drift velocity has the same direction as the force, and obliges the lines of force to go across the asymptotes of the lines of force (Fig.19).

An attempt to relate the neutral point instability to the flare phenomenon is due to Sweet (1958), who has shown in a two-dimensional model that, in the presence of a gas pressure, the neutral point can be the seat of an instability if the magnetic pressure $B^2/2\mu_o$ at some characteristic point becomes lower than the gas pressure P.

However, as an implicit idea was to associate the neutral point instability to all flare phenomena, an important question which was raised, was the question of the characteristic time of the development of the instability.

Parker (1963), discussed once again the question of the destruction of the magnetic field, which is associated with the drift velocity $\vec{w}$. The characteristic time is of the order of (L^2/η), just as in the case of the diffusion of the magnetic field; L is a characteristic length associated with the neutral point structure.

The considerable length of time needed to produce the

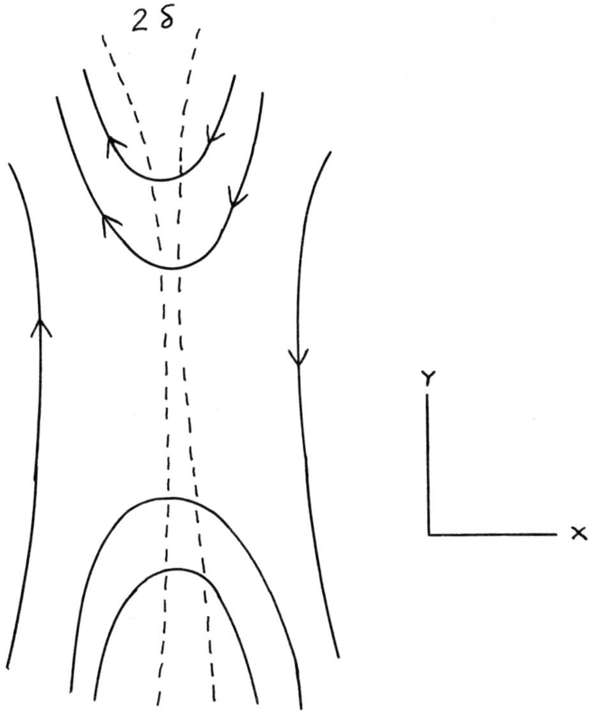

Figure 19

destruction of the magnetic field led to two different considerations: one was to assume the existence of very small characteristic lengths; the other was with Petscheck (1963) to take into account the flow of the matter near the neutral point and to show how the transfer of magnetic energy to mechanical energy might in fact considerably reduce the time scale. Let us consider, with Petscheck, a layer of thickness 2δ, the thickness being a function of the distance to the neutral point.

Far from the neutral point, wave propagation will determine the flow pattern, whereas near the neutral point, diffusion will dominate. Outside the region of strong bending of the lines of force, the magnetic field is little disturbed. The propagation of the boundary will take place at Alfvén velocity.

The boundary layer will be related to the flow velocity, towards the boundary U_{o_x} and along the boundary, v_δ:

$$U_{o_x} \quad Y = v_o \delta \tag{3.2}$$

The momentum equation in the y direction may be written as

$$\frac{d}{dy}\left(\rho v^2 \delta\right) = -\frac{B_{y_o} B_x}{\mu_o} \tag{3.3}$$

where B_x is the x component of the magnetic field within the boundary layer and is also a function of y. The above equation has equated the rate of change of momentum flux within the boundary layer to the magnetic forces.

Introducing the Alfven velocity

$$V_A = \frac{B_{y_o}}{\sqrt{\mu_o \rho}}$$

a Mach number

$$M_o = \frac{u_{x_o}}{V_A}$$

and the ratio

$$b_x = \frac{B_x}{B_{y_o}}$$

the momentum equation becomes

$$M_o^2 \frac{d}{dy} \left|\frac{y^2}{\delta}\right| = -b_x \tag{3.4}$$

Assuming that, as said above, at the boundary $|b_x| = M_o$, we obtain

$$\delta = M_o |y| \tag{3.5}$$

or the thickness of boundary layer increases linearly with y.

In the neighborhood of the neutral point, b_x varies linearly with y, as can be obtained by writing that the diffusion velocity is equal to the flow velocity:

$$b_x = \frac{-M_o^3 8\pi\sigma V_A}{c^2} y \tag{3.6}$$

The limit of the region where the diffusion takes place is obtained by writing $b_x = M_o$, which gives

$$y^* = \frac{c^2}{8\pi \sigma V_A M_o^2} \tag{3.7}$$

The perturbation $B_y \vec{\epsilon}$ of the magnetic field, where $\vec{\epsilon}$ is a vector small compared to unity can be determined in the following way. The equations of motion give $B_{y_o} \epsilon_x = B'_x$ along the boundary layer. This can be considered as magnetic sources along the boundary layer, determining $B'_y = B_{y_o} \epsilon_y$. We obtain then

$$\epsilon_y = \frac{1}{\pi} \int_{-L}^{+L} \frac{\epsilon_x (\vec{r} - \vec{\eta})}{|\vec{r} - \vec{\eta}|^2} d\eta \tag{3.8}$$

And ϵ_x is determined from the above results by the relations

$$\epsilon_x = -2M_o \frac{y}{|y|} \qquad y>y^*$$

$$\epsilon_x = -2M_o \frac{y}{y*} \qquad y<y*$$

The main contribution to ϵ_y is then

$$\epsilon_y \simeq -\frac{2M_o}{\pi} \log\left(\frac{L}{y*}\right) \tag{3.9}$$

where L is the extension of the boundary layer. The maximum value of ϵ_y compatible with the linear analysis corresponds to the maximum flow. Petscheck takes $Cy = -\frac{1}{2}$ and this gives

$$M_{o_{max}} = \frac{\pi}{4 \log (2M_{o_{max}}^2 R_m)} \tag{3.10}$$

This has to be compared to the diffusion time expressed in dimensionless units:

$$M_o = R_m^{-1/2}$$

with

$$R_m = \frac{4\pi\sigma V_A L}{c^2} \tag{3.11}$$

is the Magnetic Reynolds number.

R_m is large, due to the high conductivity of the solar plasma. The time scale is therefore comparable to (L/V_A) instead of $\frac{L}{V_A} R_m^{-1/2}$. This leads to time scales of the order of 10^2 to 10^3 seconds, instead of $5x10^4$ at least.

However, with the plasma conductivity, the thickness of the boundary layer

$$\delta = \frac{c^2}{4\pi\sigma V_A M_o} \tag{3.12}$$

is very small. At $T = 10^6$ degrees, this gives $\delta \simeq 10^{-4}$cm. This is unreasonable, as it is much smaller than the radius of gyration. With a radius of gyration of 1 Cm, the conductivity is 10^4 smaller, but this reduces the annihilation time by only 30%.

It is quite possible to substitute the conductivity of a turbulent plasma, leading to much more reasonable values of the size of the boundary layer without much changing the time of annihilation. This might well be related to the solar flare phenomenon.

E. Stimulated Flares, Sudden Disappearances, Surface Waves

The decompression wave which produces the optical flare can also propagate over large distances at the surface of the sun. Two effects can be explained by the arrival of a decompression wave:

(1) when arriving on a dark filament, the decrease of magnetic field produces a large increase of the heat flow from the surrounding corona. The excitation of the filament increases. The filament can disappear and eventually become temporarily bright;

(2) when arriving on an invisible filamentary structure in the chromosphere, the change of heat conductivity can produce a stimulated flare.

It is quite clear that the stimulated flare does not appear just by chance in a given place. Several physical conditions have to be met:

(a) the filamentary structure should pre-exist;

(b) the energy flow in the surrounding region should be large enough to provide the energy output of the optical flare;

(c) the magnetic field should be such that heat conductivity is actually modulated by decompression;

(d) heat conductivity should be high enough for the time scale of the heat transfer not to exceed 10^2 seconds.

The fast propagation (1000 km/sec) observed by Moreton (1965) at the surface of the sun can be explained also by the propagation of a wave at the surface of the sun.

Along the path of the wave, modifications of the magnetic field immediately produce a change in heat conductivity which makes brighter (decompression) or darker (compression) all the filamentary structures associated with the chromospheric pattern of the supergranulation.

F. Structure of the Magnetic Field

Before proceeding with the study of the accelerating mechanism, let us consider the structure of the magnetic field in which it takes place.

Measurements of the magnetic field have been made at photospheric level; however, it is difficult to derive the structure of the magnetic field in the chromosphere and corona from these measurements. A simple assumption would consist of continuing the photospheric field into the corona as a vacuum field. This implies surface currents in the photosphere which would produce large horizontal electromagnetic forces; in fact, there is little doubt that there are currents in the chromosphere and corona, such that the magnetic fields are very nearly force-free.

There is little doubt that the lines of forces from a north pole spot generally go to the nearby south pole spot. In that respect, the topology of the magnetic field is related to a vacuum field, but that is certainly the farthest we can go with the analogy.

Coronal condensations, above bipolar magnetic groups, show a structure which looks very much like that of a dipole field. This simply means that we definitely have tubes of force joining regions of opposite polarity.

Let us first consider an ideal picture of a bipolar group. A wide region around the sunspots shows a magnetic field, whereas in a spot, the field is of the order of 3000 gauss, and it can be of several hundred gauss over a large disturbed region. The structure of the magnetic field is influenced by the solar wind. In the case of a perfect conducting fluid, the lines of force are carried by the matter. For special models, it is possible to calculate the shape of the lines of force. An interesting result is that it takes a finite amount of time for any piece of matter to be brought to infinity.

Consider, for example, the coronal model of Van de Hulst:

$$N_e = 10^8 \frac{1.22}{r_*^{6}} + \frac{2.99}{r_*^{16}} \qquad (3.14)$$

where r_* is in solar units.

On the earth, if we observe a solar wind of 100 particles cm^{-3} at a speed of 100 km sec^{-1}, the time necessary for evaporation from the solar surface is

$$t \simeq 10^5 - 1.55 . 10^5 \left(\frac{0.41}{r^3} + \frac{0.23}{r^{13}} \right) \qquad (3.15)$$

In other words, in about 30 hours, any line of force from the lower corona is carried to infinity. Naturally it takes longer to carry away the lines of force from the chromosphere (as a simple consequence of the solar wind), but that time is of the order of one week from a low chromospheric level.

Schatzman (1965) and Sturrock (1966) have given the same representations of the shape taken by the lines of force (Fig.20). The picture of the perfect conducting fluid is certainly not valid in these regions where the lines of force are very close to each other, and of opposite direction. Due to finite conductivity of the medium, it is very likely that a neutral sheet will appear, and it has been suggested that a singular point, with a "Y" shape, would be present.* In all cases, one would expect the lines of force to be divided into two classes, the lines of force which go to infinity, and the lines of force which go from one pole to another. One observational remark must be made at this point: The observations of streamers during eclipses are a very good representation of the solar wind's effect on the lines of force of bipolar groups.

Another remark concerns the properties of the tubes of force which connect sunspots. Such tubes of force show a much larger magnetic field near the sunspots than in the corona. They can be considered to form a magnetic bottle, in which particles can be reflected from one "bottleneck" to the other.

G. The Field of Random Alfvén Waves.

We have seen that a fairly large fraction of the incident

* See for example, Green, 1965.

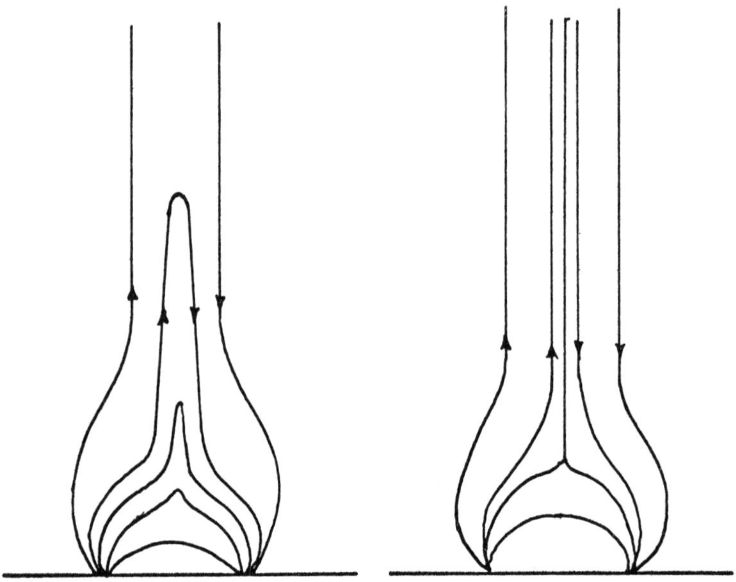

Figure 20a

flux of mechanical energy is likely to be converted into Alfvén waves. But, due to the probabilistic nature of the source and its generation through the interaction with the fluctuations of the chromosphere, the Alfvén waves produced have a random distribution of phases and amplitudes.

Alfvén waves propagate essentially along the lines of force, even in a non-uniform field, and do not refract. Therefore the decrease of mean intensity would be entirely due to dissipation.

If we consider a velocity of propagation V_A, the mean intensity will be

$$\bar{I} = \frac{1}{2} \rho <v^2> V_A \tag{3.16}$$

in which $<v^2>$ is the mean square velocity of the waves. This is also equal to the mean intensity of electromagnetic energy. The density of flowing electromagnetic energy is $\frac{1}{8\pi} (\delta B)^2$, and one finds

$$\bar{I} = \frac{1}{8\pi} (\delta B)^2 V_A \tag{3.17}$$

The electric field is related to the magnetic field and velocity by the relations

$$E = \frac{vB}{c} = V_A < \frac{\delta B}{c} > \quad . \tag{3.18}$$

At the base of the corona, for a number density of 3×10^{10} cm^{-3}, an intensity of mechanical energy of 10^6 ergs cm^{-2} sec^{-1}, and an Alfvén velocity of 10^2 km sec^{-1}, yields about 20 km sec^{-1} for the r.m.s. velocity, whereas the sound velocity is of the order of 200 km sec^{-1}. The corresponding r.m.s. magnetic field, B, is of the order of 1.4 gauss, for a local field of about 7 gauss.

Observations of both type II flares and of the propagation of perturbations along the solar surfaces give velocities of the order of 1000 km sec^{-1}. This may correspond to a higher magnetic field, and to smaller r.m.s. velocities, of the order of 0.5 km sec^{-1}.

The correlation length will be related to the correlation time by the relation

$$L = V_A \tau \tag{3.19}$$

This seems to be equivalent to considering a short characteristic time of 2 to 3 seconds, and correlation length the order of 1000 kms.

H. Motion of a Particle

Let us first consider the mean free path of a particle in the corona. We shall consider only ions.

At thermal energies, the mean free path is of the order of

$$\lambda_T \simeq \pi \frac{k^2T^2}{e^4} \frac{1}{N_e} \frac{1}{\log \Lambda} \tag{3.20}$$

in which Λ is related to the screening effect of electric charges by each other. For $N_e = 3 \times 10^{10}$ cm^{-3}, $T \simeq 2 \times 10^6$, λ_T is of the order of 1000 km.

When the velocity of the ions is much smaller than that of electrons, the energy loss is given approximately by an expression given by Gryzinski (1957),

$$\frac{dE}{d_x} = -\frac{8}{3} \frac{e^4}{m_e} \left(\frac{m_e}{8kT}\right)^{3/2} Z_A{}^2 v_A N_e. \tag{3.21}$$

The energy loss grows with the velocity, but the mean free path increases with velocity.

When the velocity of ions is much larger than that of electrons, the cross-section decreases quickly, and the loss decrease is given by

$$\frac{dE}{d_x} = -\frac{4\pi e^4}{m_e v_A{}^2} Z_A{}^2 N_e \log\frac{v_A{}^2}{\langle v_e{}^2 \rangle} \tag{3.22}$$

The maximum energy loss takes place when the velocity of ions is comparable to the thermal velocity of electrons. The mean free path then is of the order of $\lambda_m \simeq \frac{m_A}{m_e} \lambda_T$ (one million kilometers).

For higher velocities, the mean free path increases with the square of the energy.

We see from these estimates that a particle at thermal energies has a very small mean free path compared to the size of the magnetic bottle, whereas a particle of 100 kev has a mean free path comparable to the size of the magnetic bottle, and can move from one bottleneck to another.

We can attempt to describe the motion of a particle in a field of random Alfvén waves by assuming that the radius of gyration is very small, compared to the correlation length of the fluctuations, and that the period of gyration is very small, compared to the correlation time of the fluctuations.

Two effects must be considered; one is the diffusion in velocity space, and the other is that in configuration space. The diffusion in velocity space is similar to the Fermi mechanism, since the probability of a test particle having an increase in energy is greater than the probability of its having a decrease in energy. The diffusion in configuration space is due to the directional change of the particle's guiding center velocity.

This is the procedure shown by Jokipii (1966). He considers a magnetic field $\vec{B} = \vec{B}_o + \delta\vec{B}$, in which $\delta\vec{B}$ is the fluctuating part (defined by the auto correlations) and he develops the motion of the particle to the second order in $|\delta\vec{B}|$.

However, we shall follow a more elementary procedure. A diffusion coefficient in ordinary space is the product of a length by a velocity. However, we cannot take L and V for the correlation length and the velocity of the particle, but only the transverse motion of the particle over a distance along the magnetic field fluctuations and the transverse velocity, which brings the particle in the region of the next fluctuation. Clearly, we have to introduce the distance δL and the velocity δV defined by:

$$\delta L = \frac{\delta B}{B} L$$

$$\delta V = \frac{\delta B}{B} V \tag{3.23}$$

Taking into account the fact that diffusion takes place equally in directions x and y perpendicular to B, for the diffusion coefficient, we can write

$$D_{XX} = \frac{1}{2} LV \left(\frac{\delta B}{B}\right)^2 \tag{3.24}$$

This is almost exactly the expression which can be derived from the Jokipii expressions when the radius of gyration (a) is small and the gyration frequency is large.

The quantity

$$V_D = \frac{D_{xx}}{\Delta X} \qquad (3.25)$$

is a diffusion velocity. The density gradient has a scale ΔX.

If we take the orders of magnitudes used above, $(\delta B/B) \simeq 0.1, L \simeq 10^8$ cm, and $V \simeq 4.5.10^8$ cm sec^{-1} (100 kev), we obtain $D \simeq 2.2.10^{14}$ cm^2/sec^{-1}. For a density gradient of the particles with a scale of 10^5 kilometers, this gives a diffusion velocity of 0.22 km sec^{-1}. As can be expected, the diffusion velocity increases with the energy.

We can compare these orders of magnitude with the results of Bryant et al. (1965) concerning the continual acceleration of solar protons in the Mev. region.

Let us take a large solar event, with a scale of 5 proton cm^{-2} sterad^{-1} sec^{-1} at the distance of the earth, with energies $E > 2$ Mev. Such protons are confined to a stream of about 120° wide. This corresponds to a total flux of about 5×10^{27} protons coming from an active region.

If we consider leakage through a sphere of diameter d, we can take the diffusion velocity as the ratio of the diffusion coefficient to the diameter of the sphere:

$$V_D = \frac{D_{xx}}{d} \qquad (3.26)$$

If the total number of protons with energy $E > E_o$ is N_p, this gives a flux

$$\frac{Np \quad 12}{\pi d^3} \cdot \frac{\pi d^2}{2} \cdot \frac{D_{xx}}{d} = \text{flux}, \qquad (3.27)$$

or

$$d^2 = \frac{6D_{xx}Np}{\text{flux}} \qquad (3.28)$$

We take the case in which the flux is the largest of the

order of 5×10^{27} sec^{-1} $.cm^{-2}$ (angle of 120° at the earth, flux of 5 particles cm^{-2} at the earth), and

$$D_{xx} = \frac{1}{2} LV \left(\frac{\delta B}{B}\right)^2 \simeq 10^{15} \ cm^2/sec^{-1}. \tag{3.29}$$

The question remains concerning the number of particles above 2 Mev. If we assume that all these particles are accelerated during a solar flare, we can take a flux of 10^{28} and up for an average flare consisting of particles with an average energy of 100 Mev = 10^{-4} erg/particle, which gives $N_p = 10^{32}$ cm^{-2}. We then obtain

$$d^2 = \frac{6 \times 10^{15} \times 10^{32}}{3 \times 10^{27}}$$

$$d \simeq 1.4 \times 10^{10} \ cm$$

This is a reasonable size for the leaking sphere. The corresponding diffusion velocity is v_D = 0.7 km/sec. We can consider that this is a remarkable agreement, and we can conclude that a continual acceleration takes place at the surface of the sun.

Let us now consider the diffusion in velocity space. The change of energy of the particles is essentially due to the stochastic electric field associated with the random Alfvén waves. There is a net energy gain due to the fact that the energy, on the average, is larger than the energy loss, causing a diffusion in the velocity space.

We can say that the longitudinal accelerating field is in the ratio of $\frac{a}{L}$ to the transverse electric field $\frac{vxB}{c}$. In the same way, we can consider that the distance of acceleration is like L $\frac{\delta B}{B}$. We can therefore obtain the energy gain over a correlation length L,

$$\Delta E/E \simeq e \ \frac{a}{L} \ \frac{\delta B V_A}{c} \ . \ L \ \frac{\delta B}{B}, \tag{3.30}$$

or

$$\frac{\Delta E}{\Delta X} \simeq m_A V_A{}^2 < \left(\frac{\delta B}{B}\right)^2 > \frac{1}{L} \tag{3.31}$$

If we want to express that the random field of Alfvén waves can accelerate the particles, we have to denote that the maximum losses are smaller than the energy gains. This can be written

$$m_A V_A^2 < \left(\frac{\delta B}{B}\right)^2 > \frac{1}{L} > \frac{8\pi}{3} \frac{e^4}{m_e} \left(\frac{\pi me}{8kT}\right)^{3/2} v_e N_e \tag{3.32}$$

We try to eliminate unknown quantities and use only known quantities like L, the correlations length; $\overline{I}$, the mean intensity of mechanical energy; v_A, the Alfvén velocity; and T, the temperature of the corona. This gives

$$N_e < \frac{24}{\pi^2} \frac{kT}{e^4} \frac{\overline{I}}{LV_A}$$

for a fully ionized hydrogen gas, or numerically,

$$N_e^2 < 10^{21.8} \frac{T\,\overline{I}}{LV_A} \tag{3.33}$$

For $L = 10^8$ cm, $V_A \simeq 10^8$ cm sec^{-1}, as deduced from the velocity of propagation of the type II bursts, $\overline{I} = 10^5$ ergs cm^{-2} sec^{-1}, an acceptable value at a high coronal level, $T = 2 \times 10^6$ °K, we have $N_e < 4 \times 10^8$ cm^{-3}.

The condition is fulfilled in the corona. Therefore it seems that continuous pumping of energy can take place out of the Maxwell distribution function of the ions.

I. Accelerating Mechanisms.

The random field of Alfvén waves cannot accelerate particles beyond the Alfvén velocity; another mechanism has to carry the particles to higher energies. It has been suggested by Tsytovitch (1966) that the appropriate mechanism is plasma turbulence.

We shall say that there is a plasma turbulence when the energy density of the plasma waves is higher than that at thermodynamic equilibrium.

At thermodynamic equilibrium, the energy included in these collective modes is related to the energy included in the individual modes in the following way: If Λ_o is the Debye length, and N_e the number density of the electrons, then the ratio of the collective energy, E_{coll}, to the energy of individual particles, E_{ind}, is:

$$\frac{E_{coll}}{E_{ind}} = \frac{1}{\frac{4}{3}\pi \Lambda_o^3 N_e} \tag{3.34}$$

This is usually a small quantity, which means that in ordinary plasmas, the energy density of the fluctuating electric field is small, compared to the thermal energy of the particles. In the corona, it is of the order of 10^{-8}.

According to Tsytovitch, it seems that plasma turbulence is easily excited at low densities, and it is suggested here that the random field of Alfvén waves excites the plasma turbulence. If again equipartition is obtained, this means that an appreciable fraction of thermal energy goes into plasma turbulence.

If we take a mean intensity $\bar{I}$, we write

$$\frac{1}{2}\rho v^2 V_A = \bar{I} \tag{3.35}$$

and if we divide it equally in the collective modes, this gives

$$\frac{E_{coll}}{E_{therm}} = \frac{\bar{I}}{3V_A N_e kT} \tag{3.36}$$

For $\bar{I} = 10^5$ erg cm sec^{-1}, $V_A = 10^8$ cm/sec^{-1}, $N_e = 10^8$ cm^{-3}, and $T = 2.10^6\ {}^oK$, this gives

$$\frac{E_{coll}}{E_{therm}} = 10^{-2} \tag{3.36'}$$

The corresponding electric field is about 1000 times larger than the field at equilibrium.

Acceleration by a stochastic electric field can take place when the velocity of the protons is larger than the thermal velocity of the electrons; in other words, the velocity of the accelerated particle has to be larger than the phase velocity of the plasma waves. However, according to Tsytovitch, it seems that in the presence of a magnetic field, the low-phase velocity of transverse plasma waves permits transverse acceleration to take place. This might bridge the gap between the acceleration by the random field of Alfvén waves and plasma turbulence acceleration.

A more correct analysis can be carried by using the first order smoothing approximation, the reader can be referred to the lectures given at Les Houches in 1968.

J. Solar Cosmic Rays.

We have seen in the last chapter that continuous acceleration by a field of random Alfvén waves can take place in the magnetic bottle existing in the neighborhood of sunspots.

If we calculate the energy gain over one correlation length, we obtain (using the expression of the last chapter): $\langle \Delta E \rangle \simeq 10$ e.v. This gives the order of magnitude 10^5 seconds for the time necessary to reach 10^6 e.v., if we assume that this is a constant during acceleration. When a bipolar magnetic group develops, it takes about 5 days until the first flares appear in the group, according to Kiepenheuer (1953). From the order of magnitude of the characteristic time for the growth of the particle's energy, we surmise that the magnetic bottle which appears after 2 to 3 days must first be filled with low-energy cosmic ray particles before the flares can appear.

We can expect that the accelerating mechanism will fill the magnetic bottle with low-energy cosmic rays up to the point where the energy density of the cosmic ray is comparable to the energy density of the magnetic field. In a volume of 10^{30} - 10^{31} cm^3, this corresponds to a total number of particles of the order 10^{35}, and a total energy of the magnetic field of the order 10^{31} to 10^{33} ergs.

We shall consider now two possible explanations for the acceleration to energies of the order of 1 Gev.

The first explanation is due to Syrovatskii (1966). It is similar to the mechanism of Petschek. However, there is an important difference. In the Petschek mechanism, there is conversion of magnetic energy into mechanical motion; in the Syrovatskii mechanism, there is conversion of mechanical motion into particle motion through the formation of a large electric field.

Syrovatskii considers a neutral point of the magnetic field which is compressed by the motion of the regions generating the magnetic field. Near the neutral point, Syrovatskii considers two regions, one in which the motion of the magnetic field determines the motion of the matter, another in which the matter is compressed and carries the magnetic field.

The vector potential due to two current-carrying wires placed at $x = \pm 1$ is

$$A = \frac{h_o}{4} \log \left\{ \left[(x-1)^2 + y^2 \right] + \log \left[(x+1)^2 + y^2 \right] \right\} \tag{3.37}$$

Near the neutral point,

$$A = \frac{h_o}{2} (y^2 - x^2) \tag{3.38}$$

Syrovatskii assumes that the two currents are displaced by a quantity δ. The density is changed according to the change of magnetic field:

$$\frac{\rho(x,y)}{\rho_o} = \frac{x^2 + y^2}{\sqrt{(x^2 + y^2) + 4\delta(x^2 - y^2) + 4\delta^2}} \tag{3.39}$$

If $y^2 < x^2 + \delta$ there is rarefaction, if $y^2 > x^2 + \delta$ there is compression. Finally, near the neutral point, there is a region of radius r_s,

$$r_s^2 = \frac{4\pi\rho_o s^2}{h_o^2} \tag{3.40}$$

where s is the sound velocity, inside which the motion of the matter carries the magnetic field.

The lines of force are carried by the motion of the currents, but, near the neutral point, remain frozen in (Fig. 20 b). The magnetic field increases in the central region, and takes the value h.

The current near the neutral point is given by $\frac{1}{\mu_o} (\nabla \times B)$. For the vacuum field, with $B_x = h_o y$, $B_y = h_o x$, $\nabla x B = 0$; but, when the field is compressed, $B_x = h_1 y$, $B_y = h_2 x$, $h_2 \gg h_1$ and the current is given by

$$j_z = h_2 - h_1 \simeq h_2$$

The coefficient h_2 can be estimated by considering the magnetic field at the border of the compressed region.

The field was $h_o \sqrt{\delta}$; the flux was $\frac{1}{2} h_o \delta$ along the x axis; this has been compressed to r_s, and the new flux is of the order of $\frac{1}{2} h_2 r_s^2$ from which we derive

$$h_2 \simeq \frac{h_o \delta}{r_s^2} \tag{3.41}$$

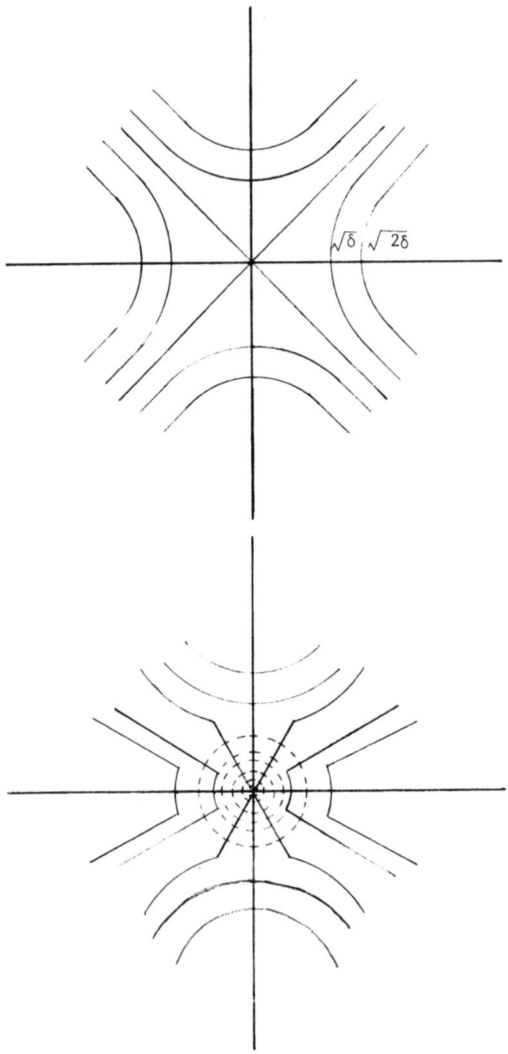

Figure 20b

At the beginning, the current is still given by $\sigma \times B$, and is associated with an electric field E,

$$E = \frac{j}{\sigma} = \frac{ch}{4\pi\sigma}$$

Syrovatskii writes simply that the electric field is large enough to produce runaway particles. This gives

$$h > \frac{4\pi\sigma \ Erit}{c} = \frac{4\pi \left(\frac{3\sqrt{3}}{4\pi L e^2 \sqrt{m}}\right)(kJ)^{3/2}}{c} \quad \frac{4\pi e^3 Ln}{kT} \tag{3.42}$$

where L is the Coulomb logarithm.

It is easy to show that all the matter which was initially in a cylinder of radius $\sqrt{2\delta}$ is submitted to such an electric field. This gives the possibility of estimating the energy available per unit length,

$$W_m = \frac{1}{8} h^2 \delta^2 \tag{3.43}$$

and the number of accelerated particles,

$$N = 2n_o \int_{r_s}^{\sqrt{2\delta}} \frac{r_s^2}{x} dx = 2n_o r_s^2 \log \frac{\sqrt{2\delta}}{r_s} \tag{3.44}$$

The effect takes place if δ is large enough, and if the time scale of the compression is short enough:

$$\tau << a\frac{\sqrt{\delta}}{s}$$

where a is the scale of the system, and s the sound velocity.

The average energy per particle is then

$$\epsilon = \frac{h_o r_s^2 \ 4\pi e \left(\frac{T}{5.10^8}\right)^{1/2} a^2}{16 \log \frac{\sqrt{2\delta}}{r_s}} \tag{3.45}$$

(with $\frac{mc^2}{k} = 5.10^8$ degrees)

Orders of magnitude are in agreement with solar cosmic ray

observations. With

$$n_o = 10^9, \quad T = 10^6, \quad a = 10^{10}\text{cm}$$

$$s = 10^7\text{cm/sec}$$

$$h_o = 10^{-6} \text{ gauss cm}^{-1}$$

$$r_s^{\,2} = 3.10^{-10} \quad (ar_s = 2.10^{-5}a = 2.10^5\text{cm})$$

The condition for δ is

$$\delta^2 \gtrsim \frac{1}{2} \cdot 10^{-8} \qquad\qquad a\delta \gtrsim 10^{-4}a \simeq 10^6\text{cm}$$

The characteristic time of the process and the displacement rate of the spots should be $\tau \ll 10$ sec and $v \gg 10^5$ cm/sec. The energy converted into fast particles is

$$W_m \simeq 10^{29}\text{ergs}$$

The number of accelerated particles

$$N \simeq 10^{30}$$

and the mean particle energy 10^{11}e.v.
With other numbers, a satisfactory agreement can be obtained (10^{31}ergs, 10^{34} particles, 10^9e.v.)

We shall assume now that the solar flare is due to a release of energy which accelerates further the low-energy cosmic ray particles and brings them to energies of the order of 100 Mev to 1 Gev.

Let us first consider the energy released in solar cosmic rays. According to McDonald (1963), the 3+ flare of July 16, 1959 was accompanied by an integrated intensity of 8.1×10^8 particles cm^{-2} for particles of energy $E > 30$Mev. This corresponds to an energy in one steradian of about 2.10^{31} ergs for a number of about 2.10^{34} particles in one steradian.

We shall assume that the energy release is due to an opening of the magnetic bottle which simultaneously accelerates the low energy C.R. particles which it contains and throws them away in interplanetary space. The mass motion of the magnetic bottle induces a stochastic electric field which, over a correlation length, gives energies of the order of $\frac{1}{2}$ Gev. We can assume that the acceleration mechanism pumps out of the low energy C.R. particles to produce higher energy C.R. particles, with the acceleration taking place only during the very short time of the opening of the magnetic bottle, that is, during a time smaller than one hour. It seems reasonable to apply to that phase of the acceleration the thermodynamic reasoning of Syrovatskii (1961), which leads to a differential power law with an exponent belonging to the interval 3.5-5.

The expansion of the magnetic bottle, as a result of an instability, is probably the general mechanism of the solar flare. The appearance of C.R. at the Earth can be due to the opening of the magnetic bottle, whereas in other cases, a release of energy can take place without the opening of the magnetic bottle.

K. Triggering of the Optical Flare.

The opening of the magnetic bottle is accompanied by the production of an expansion wave which propagates across the lines of force of the magnetic field. If we assume that the instability takes place in the corona at the altitude where the type III and type II bursts intersect, i.e., at an altitude of 50,000 to 100,000 kilometers, it will take about 30 to 100 seconds for the disturbance to reach the region of the filamentary structure where it triggers the optical flare.

Let us again consider the 3^+ flare of July 16, 1959. If we assume the instability taking place at an altitude of 100,000 kilometers, we find that the expansion wave has reached all the appropriate filamentary structures in a time of the order of 3 minutes, whereas the observations indicate that the commencement took place from 21:19 to 21:24 UT. i.e., in 5 minutes. If we include the relaxation time of the heat transfer from the corona, we are quite well inside the limits imposed by the observations.

The instability itself can be due to the classical neutral-line instability. However, instead of assuming that the neutral-line instability takes place at the level of the optical flare and that the destruction of the magnetic field has to provide all the energy of the solar flare, we shall assume that the neutral-line sits at a high altitude (50,000 to 100,000 kilometers) in the corona, and that the instability does not provide the liberation of the energy of the flare, but just has the role of a triggering mechanism. Annihilation of the magnetic field taking place locally could well trigger the change of shape of a magnetic bottle filled with low energy C.R. particles. The presence of plasma turbulence, as assumed by Syrovatskii (1966) or by Wentzel (1964) and Jaggi (1964) could well explain the short time scale of the magnetic field annihilation, as a consequence of the resistivity increase.

The propagation downwards of the expansion wave could also explain the propagation downwards of the excitation observed by Severny (1964).

L. Summary and Conclusion.

The picture of solar flares which is proposed here assumes essentially that an instability taking place in the high corona can trigger the optical flare and the accelerating mechanism which produces solar C.R. particles in the 100 Mev range.

The optical flare consists essentially of a change of the heat balance of filamentary structures, pre-existing to the flare. The change of heat balance is induced by a decrease of the magnetic field due to the arrival of a decompression wave. The energy involved in the optical flare is carried by thermal conductivity from the corona to the filamentary structure. The energy of the flare is supplied by the flux of mechanical energy which enters in the corona and comes from the hydrogen convective zone of the sun. The place of appearance of the X-ray flashes and the 3 mm radio-waves should be located near the sunspots, where the fast electrons channeled by the magnetic field could hit the matter of the low chromosphere. The place of commencement of the optical flare is determined by the arrival of the expansion wave at the place where the physical conditions in the filamentary structure are the proper ones.

A final question can be raised now: what is the interpretation of the changes in the magnetic field observed by Howard and Severny (1963) at the photospheric level? The answer to be given here is that it does not correspond to a destruction of the magnetic field but only to a mass motion of a small amount of matter which, at the photospheric level, brings magnetic fields which were hidden before under the photosphere. The energy involved then corresponds only to the magnetic energy stored inside the observable region of the photosphere, which is only 10^7 cm thick. The energy change is then at most 4×10^{30} ergs, and then much smaller than the energy radiated by the optical flare.

Detailed considerations of a flare studied by de Jager (1967) show the possibility of a full agreement between the suggested theory of the solar flares and the observations.

As an order of magnitude, we can compare the time necessary for the fast electron to produce X-rays and the time necessary for the beginning of the optical flare.

We can assume that the electrons are trapped in a magnetic bottle which moves with Alfvén velocity along the lines of force until it arrives in a region of sufficiently high density for the bremsstrahlung to take place.

On the other hand, the optical flare will take place after the expansion wave has reached the chromospheric level.

As an order of magnitude, we can assume that the expansion wave propagates at the velocity $\sqrt{2V_A}$, and that the traveling distance is d for the expansion wave and $\frac{\pi}{2}$ d for the Alfvén wave.

After the triggering of the optical flare, we would have the following log:

X-rays: $\frac{\pi d}{2V_A}$

optics: $\frac{d}{\sqrt{2V_A}}$ + heat relaxation time .

With $d \simeq 10^5$ kms and $V_A \simeq 1500$ km, we get

$t_{X\text{-rays}} \simeq 150$ seconds

$t_{op.\ fl} \simeq 50$ seconds $+ t_{HRT}$.

As the heat relaxation time is of the order of 100 seconds, we see that the X-ray flash can coincide with the optical flare, but that there is no reason for exact coincidence.

BIBLIOGRAPHY

Chapter III

ASSEO, E., and BERTHOMIEU, G., (1966), "Accélération de particules chargeés par un champ électromagnetique aléatoire," Ann. d'Ap., 29, 689.

ATHAY, R. C., and THOMAS, R. N., (1961), "Physics of the Solar Chromosphere," New York, Interscience Publishers, Inc.

BECKERS, J. M., (1963), "Study of the Undisturbed Chromosphere from Hα-Disk Filtergrams, with Particular Reference to the Identification of Spicules," Ap. J., 138, 648.

BERNAS, R., GRADSZTAJN, E., REEVES, H., YIOU, F., and SCHATZMAN, E., (1967), "The Nucleosynthesis of Lithium, Beryllium, and Boron, and the Formation of the Solar System," Annals of Physics, 44, 426.

BONSACK, W. K., and GREENSTEIN, J. L., (1960), "The Abundance of Lithium in T Tauri Stars and Related Objects," Ap. J., 131, 83.

BRYANT, D. A., CLINE, T. L., DESAI, U. D., and McDONALD, F. B., (1965), "The Continual Acceleration of Solar Protons in the MeV Region," Goddard X-611-65-107.

De FEITER, L. D., (1963), "On the Stark Broadening of High Balmer Lines Emitted by Solar Flares," 1963 AAS NASA Sympsoium on Solar Flares, 81.

De JAGER, C., (1967), "The Hard Solar X-Ray Burst on September 18, 1963," Solar Physics, 1, 327.

DOHERTY, L. R., and MENZEL, D. H., (1965), "Filamentary Structure in Solar Prominences," Ap. J., 141, 251.

DUNGEY, J. W., (1958), Cosmic Electrodynamics, Cambridge University Press.

GIOVANELLI, R. G., (1947), "Magnetic and Electric Phenomena in the Sun's Atmosphere Associated with Sunspots," M. N., 107, 338.

GREEN, R. M., (1965), "Modes of Annihilation and Reconnection of Magnetic Fields, I. A. U. Symposium 22, 398.

GRYZINSKI, M., (1957), "Stopping Power of a Medium for Heavy Charged Particles," Phys. Rev., 107, 1471.

HARO, G., (1960), "On the Problem of the T Tauri and UV CETI Stars," Symposium on Stellar Evolution, La Plata, p. 37.

HERBIG, G., (1960), "On the Nature and Origin of the T Tauri Stars," I. A. U. Symposium No. 3, Non-Stable Stars, p. 3.

HERBIG, G., (1957), "Stars of Types F, G, and K with Emission Lines," Introductory Report, Stars with Emission Lines, 8th Liege Colloquium, p. 251.

HIRAYAMA, T., (1961), "Physical Structure of Limb Flares," Publ. Astr. Soc. Japan, 13, 152.

HOWARD, R., and BABCOCK, H. W., (1960), "Magnetic Fields Associated with the Solar Flare of July 16, 1959," Ap. J., 132, 218.

HOWARD, R., and SEVERNY, A., (1963), "Solar Magnetic Fields and the Great Flare of July 16, 1959," Ap. J., 137, 1242.

JAGGI, R. K., (1969), "A Mechanism for the Dissipation of the Magnetic Field in Solar Flares," AAS NASA Symposium of the Physics of Solar Flares, p. 419.

JOKIPII, J. R., (1966), "Cosmic-Ray Propagation, I. Charged Particles in a Random Magnetic Field," Ap. J., 146, 480.

JOY, A., (1957), "Flare Spectra in Dwarf Stars," I. A. U. Symposium No. 3, Non-Stable Stars, Cambridge University Press, p. 31.

JOY, A., and HUMASON, M., (1949), "Observations of the Faint Dwarf Star L726-8," P.A.S.P., 61, 133.

KIEPENHEUER, K. O., (1953), The Sun, Kuiper, G. P., Editor, The University of Chicago Press, p. 322.

LOVELL, B., (1964), "Radio Stars in the Galaxy," Observatory, 84, 191.

McDONALD, F. B., (1963), "Solar Proton Manual," NASA Technical Report TRR 169.

MORETON, G. E., (1965), Film shown on Dec. 3, 1965, Meudon Observatory.

ORALL, F. Q., and ZIRKER, J. B., (1961), "Heat Conduction and the Fine Structure of Solar Prominences, I. Optically Thin Model Prominences," Ap. J., 134, 72.

OSKANJAN, V., (1953), "Variations d'eclat de l'étoile UV CETI," Bull. Obs. Ast. Beograd, 18, 24.

PARKER, E. N., (1963), "The Solar Flow Phenomenon and the Theory of Reconnection and Annihilation of Magnetic Fields," Ap. J. Suppl., VIII, 177.

PETSCHECK, A. E., (1963),"Magnetic Field Annihilation," NASA Symposium on the Physics of Solar Flares, 425.

SCHATZMAN, E., (1949), "The Heating of the Solar Corona and Chromosphere," Ann. d'Ap., 12, 203.

SCHATZMAN, E., (1958), "Discussion des communications 28A 32," Stars with Emission Lines, 8th Liège Colloquium, p. 289.

SCHATZMAN, E. (1959), "On the Possibility of Observing Radio Emission from Flare Stars," IAU Symposium No. 9, Radio Astronomy, p. 552.

SCHATZMAN, E., (1965), "Particle and Radio Emission from the Sun," The Solar Spectrum, edited by C. de Jager, p. 313.

SCHATZMAN, E., (1966), "Contribution to the Theory of Solar Eruptions," C. R. Acad. Sci. B., 262, 93.

SCHATZMAN, E., (1968), Université de Grenoble, Ecole d'été de Physique Théorique, Les Houches, Summer course held in 1966, Volume III, Accelerating Mechanisms.

SEVERNY, A., (1964), "Solar Flares," Annual Review of Astronomy and Astrophysics, 2, 363.

SEVERNY, A., (1966), "Private Communication."

STURROCK, P., (1966), Cleveland Colloquium on Collisionless Shocks and Plasma Turbulence.

SUEMOTO, Z. and HIEI, E., (1959), "Balmer Series Lines of the Flare and its Structure," Publ. Astr. Soc. Japan, 11, 185.

SVETSKA, Z., (1962), "Evidence of Stark Broadening of Balmer Lines in Flares," Bull. Astr. Inst. Czech., 13, 236.

SVETSKA, Z., (1965), "Spectral Analysis of Solar Flares," Adv. Ast. and Ap., 3, 119.

SWEET, P. A., (1958), "The Neutral Point Theory of Solar Flares, IAU Symposium Number 6, Cambridge University Press.

SYROVATSKII, S. I., (1961), "Spectrum of Galactic and Solar Cosmic Rays," J.E.T.P., 13, 1257.

SYROVATSKII, S. I., (1966), "Dynamic Dissipation of a Magnetic Field," A. J. U.S.S.R., 10, 270.

TSYTOVITCH, V. N., (1966), "Statistical Acceleration of Particles in a Turbulent Plasma," Soviet Physics USPEKHI, 9, 370.

WENTZEL, D. G., (1963), "The Origin of Solar Flares and the Acceleration of Charged Particles," AAS-NASA Symposium on Solar Flares, p. 397.

IV. SURFACE NUCLEAR REACTIONS AND ACCELERATING MECHANISMS

A. Stellar Abundances and Peculiar Abundances

It is well known that in a large number of stars, the abundance of elements deviates from the so-called cosmic abundance which is determined either from the chemical composition of meteorites or from a spectral analysis of the sun and normal stars. It is usually assumed that stars are formed from interstellar matter. Due to the relatively fast mixing which takes place in interstellar matter, it is probably not a bad approximation to suppose that stars formed in a certain neighborhood during a certain interval of time are of the same composition, even though the composition can change with time.

If this is assumed, then all departures from normal composition can be attributed to the following effects:

1. Formation at different epochs. The secular changes in composition of interstellar matter can explain the difference in chemical composition of old and new stars;
2. Thermonuclear reactions in stellar interiors. This can explain the differences in chemical composition which are observed at a late stage of stellar evolution, as the overabundance of carbon in carbon stars;
3. Surface nuclear reactions. This can explain differences in chemical composition which are observed at an early stage of stellar evolution.
4. Gravitational diffusion

This process has been considered by Aller and Chapmann (1963). It can take place at the bottom of the hydrogen convective zone. For diffusion of test particles in ionized hydrogen the velocity of diffusion is

$$w_2 = D_{12}\left[-\frac{1}{c_2}\frac{\partial c_2}{\partial z} + \frac{2W - Z - 1}{P}\frac{\partial P}{\partial z} + \frac{2.65\ fZ^2 + 0.805\ (W - Z)}{T}\frac{\partial T}{\partial z}\right]$$

with the following meaning of the symbols:

D_{12} is the diffusion coefficient of element 2 in hydrogen

$W = \frac{m_2}{m_1}$, is the ratio of the mass of the diffusing element to the mass of the hydrogen atom

Z is the effective charge of the diffusing element

The diffusion coefficient is

$$D_{12} = 1.947 \times 10^9 \; T^{5/2} \left[n_1 M_2 Z^2 A_1 (2)\right]^{-1}$$

with the following meaning of the symbols:

T = absolute temperature

n_1 = number density of the hydrogen atoms

$M_2 = [w/ (1 + w)]$

$A_1(2) = \log_e (1 + x_D)^2$

$x_D = \frac{4d_D \; kT}{Z \; e^2}$; d_D, Debye length

We can call this diffusion gravitational diffusion if the pressure gradient is due to gravity.

In the chromosphere, the main effect will be the temperature gradient, which will drive the diffusion.

5. Thermal diffusion

In the processes (4) and (5), there is a competition between the driving forces (gravity, partial separation of the charges due to gravity, change of mobility with temperature) and the effect of the cross section, which increases like the square of the effective charge.

Due to the difficulty of interpreting the observations, we shall limit ourselves to the light elements. Specifically, let us consider the lithium problem. The abundance of lithium in F and G main sequence stars varies from star to star by a factor, the maximum of which is about 25. T Tau stars, which are young stars, have an abundance of lithium which is about 100 times the solar abundance.

Herbig, Wallerstein, and Conti (1965) measured the lithium abundance for approximately 100 field stars from F5 to G8, for 23 stars in the Hyades. Danziger and Conti (1966) did the same work for 69 field stars from F0 to F8 and six stars in the Pleiades.

In the Hyades, there is a drop in lithium content at spectral type G5 (Fig. 21, a, b, c), whereas in the Pleiades, the lithium content seems to be the same for F and G stars, similarly in the field

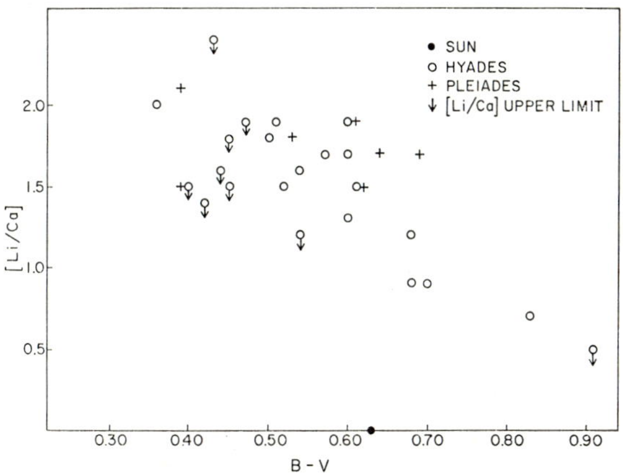

FIG. 21a–Li/Ca plotted as a function of $B - V$. *Crosses,* Pleiades stars; *circles,* Hyades stars (see text).

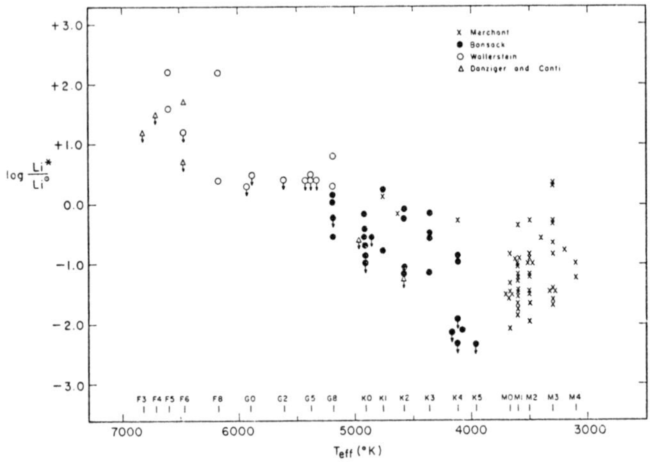

FIG.21b–The logarithmic ratio of stellar to solar Li abundance plotted against spectral type or effective temperature. The comparison element (Ca or V) was assumed to be equal to the star and the Sun. The values of Merchant and of Danziger and Conti (1966) are on the same scale; Wallerstein's (1966*b*) values were all reduced by –0.3 in the log to adjust them to the Greenstein-Richardson solar abundance of Li (see Table 5). The calibration of T_{eff} and spectral type was taken from Johnson (1964) for luminosity class III or was interpolated between Johnson's values for class V and Kraft's (1961) values for class I.

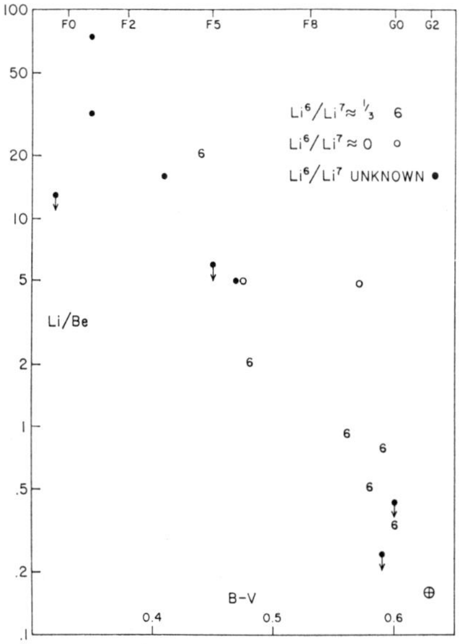

FIG. 21c–Li/Be ratios as a function of spectral type for the main-sequence stars discussed here and by Merchant (1966). An obvious correlation exists, the implications of which are discussed in the text.

stars, Herbig (1965) found a drop at spectral type G3, whereas Danziger and Conti (1966) found a loose correlation between age and lithium content, with the younger stars having the higher content.

The abundance of lithium in K dwarf stars is very small (less than 1/10 of the solar value) according to Bonsack (1959). Furthermore, Herbig measured the ratio of Li^6 to Li^7 in 15 main sequence field stars F5V to G8V. The ratio varies from 0 to 0.6. A surprising feature is the low ratio (of about zero) found by Merchant, Bodenheimer and Wallerstein (1966) in a Hyades star, 45 Tau.

Beryllium was observed by Merchant (1966) in a number of F and G dwarfs. The Be abundance varies by a factor 1000. Recalibration of the abundance of Beryllium in the Sun led Conti (1969) to estimate the (Li/Be) ratio from from 600 for FOV stars to 0.1 for a late G2V like the Sun.

B. Surface Nuclear Reactions.

The presence of light elements in stellar atmospheres can be explained only by surface formation.

Lithium, for example, is burnt in a short time, as soon as the temperature goes above $2.4\ 10^6$ degrees. Consequently, the galactic content in lithium is likely to have been continuously decreasing. Bernas, Gradsztajn, Reeves and Schatzman (1967) have assumed that the initial abundance of lithium is very small, and that cosmic rays were unable to build in interstellar space more than $(Li)/(H) \simeq 3.10^{-12}$. We therefore first suppose that lithium was formed at the surface of stars at an early stage of stellar evolution.

The following reactions can take place:

-neutron formation from C.N.O.:

$$O^{16}\ (p,n)\ F^{16}\ (\beta,\ \nu)\ O^{16}$$

-spallation from C.N.O.:

$$C^{12}\ (p,\ \alpha pn)\ Li^7$$

$$p(C^{12},\ \alpha nn)\ Be^7\ (\beta\nu)\ Li^7;$$

-thermonuclear-like reactions at Mev energies:

$$Li^7\ (p,\ \alpha)\ He^4;$$

with an energy spectrum in $E^{-\gamma}$ and a cross-section varying, as $E^{-1} \exp - (E_o/E)^{1/2}$; this has a sharp maximum for $E = 4\ (\gamma + 1)^2 E_o$.

For lithium, E_o = 7.700 kev. Cross-sections for spallation of (CNO) have the following order of magnitude:

$$Li^7 \simeq 20 \text{ mb}$$

$$Li^6 \simeq 10 \text{ mb}$$

$$Be^9 \simeq 2 \text{ mb}$$

The most difficult case is that of Be^9, which has a maximum abundance $(H)/(Be) \simeq 10^9$.

Spallation takes place in competition with bremsstrahlung. The energy necessary in building a Be atom is given by the following average:

$$< \frac{\sigma_{stop}[H]}{\sigma_{spall}[CNO]} E > \simeq 10^{13} \text{ e.v.}$$

This value is a minimum (Reeves, 1966).

C. Energy Problems.

The formation of one beryllium nucleus, by spallation of C^{12} or N^{14}, or O^{16}, requires a large amount of energy. As mentioned above, Audouze, Gradsztajn and Reeves (1967) estimate that at least 10^{13} e.v. is needed per Be formed.

It should be noticed that most of that energy is used for heating the gas in which fast particles are formed. This is due to the very low cross-section for Be formation.

In all former considerations, it was assumed, that all that energy was lost by the star, and this led to difficulties with the problem of the star's energy balance.

Let us explain these difficulties by considering the most extreme case of complete mixing inside the star. In this case, the total number of Be atoms found in a star of one solar mass is of the order 10^{48}, which represents an energy of the order of 10^{49} ergs. With a lifetime of the order of 10^7 years, this represents a cosmic ray luminosity of the order of $3 \ 10^{34}$ ergs/sec. This high value is not acceptable, as it is impossible to imagine an accelerating process in which the energy placed into the accelerated particles would be of the order of magnitude of the entire available energy. Another way of expressing this is by considering the energy needed by an H atom in the star to build a Be atom. This is of the order of 10^4 ev/nucleon; however, after total contraction, a star can liberate a maximum energy of about 1 kev/nucleon.

Even if we assume that the elements have been diluted in a small convective zone, we must remember that a mass equal to one solar mass is entirely convective until it contracts to a radius of 3 R. Several assumptions have to be considered: (1) diffusion in the convective zone was not too efficient; (2) spallation takes place in a hot plasma, with a small stopping power, or (3) production of light elements has taken place in interstellar space.

D. Turbulent Diffusion. I.

The equation of transport can be established in the following way: if ρ is the density, and X is the concentration of the test particles (of such low concentration that they do not affect the density), then the equation of continuity in one dimension is:

$$\frac{\partial}{\partial t}\rho X + \frac{\partial}{\partial z}\rho v X = 0 \tag{4.1}$$

We suppose that the mixing process takes place in such a way that, at the place of mixing, the concentration X is given by:

$$X = \langle x \rangle + \ell \frac{\partial \langle x \rangle}{\partial z} \tag{4.2}$$

in which ℓ is the vertical mixing length. We shall take the average of the continuity equations, assuming that $\langle v \rangle = 0$, $\langle \ell \rangle = 0$, and $\langle \ell v \rangle = - D \neq 0$. D is the coefficient of diffusion. We then obtain the equation of transport:

$$\rho \frac{\partial \langle x \rangle}{\partial t} - \frac{\partial}{\partial z}\rho D \frac{\partial \langle x \rangle}{\partial z} = 0 \tag{4.3}$$

We shall consider the solution of the diffusion equation with the following model:

(a) The distribution of density is described by a scale height, H^{-1},

$$\rho = \rho_o \text{ exo } (-Hz);$$

(b) The diffusion takes place from the level of formation of the elements z = 0 to a certain level z = -a, which is the bottom of the convective zone. In fact, we can consider that ρ_o exp (Ha) is the mass of the convective zone per unit surface;

(c) At level z = 0, we integrate equation (4.3) between -E and +E, and write:

$$\int_{-\epsilon}^{+\epsilon} \frac{\partial}{\partial z} \rho D \frac{\partial \langle x \rangle}{\partial z} dz = \rho_o \int_{-\epsilon}^{+\epsilon} \frac{\partial \langle x \rangle}{\partial t} dz \qquad (4.4)$$

calling Φ the number of nuclei formed per square centimeter and per second, and n_H the number density at level z = 0,

$$n_{H_o} \; D_o \left[\left(\frac{\partial \langle x \rangle}{\partial z} \right)_+ - \left(\frac{\partial \langle x \rangle}{\partial} \right)_- \right] = \Phi \qquad (4.5)$$

We shall assume equal partition between upward and downward diffusion:

$$\left(\frac{\partial \langle x \rangle}{\partial z} \right)_+ = - \left(\frac{\partial \langle x \rangle}{\partial z} \right)_- = \frac{\Phi}{2 n_{H_o} D} \; ; \qquad (4.6)$$

(d) At level z = -a, we shall assume that diffusion stops, the level acts as a reflector, and the elements accumulate in the convective zone;

(e) The diffusion coefficient D is taken as a constant. As we shall see, the result is independent of whether or not D is depth dependent.

With assumptions (a) and (e), the diffusion equation becomes:

$$\frac{\partial \langle x \rangle}{\partial t} + DH \frac{\partial \langle x \rangle}{\partial z} - D \frac{\partial^2 \langle x \rangle}{\partial z^2} = 0 \qquad (4.7)$$

We solve Eq. (4.7) by the standard method described by Morse and Feshbach (1953). We consider a harmonic solution of the form y(z) exp $i\omega t$. Using the boundary conditions,

$$z = -a, \qquad \frac{\partial \langle x \rangle}{\partial z} = 0,$$

$$z = 0, \qquad \frac{\partial \langle x \rangle}{\partial z} = h \exp i\omega t$$

we find

$$\langle x \rangle = \frac{h}{\exp(r_1 a) - \exp(r_2 a)} \left[\frac{\exp[r_1 (a+z)]}{r_1} - \frac{\exp[r_2 (a+z)]}{r_2} \right] \qquad (4.8)$$

with

$$r_1 = \frac{DH + \sqrt{D^2H^2 + 4\ i\omega D}}{2D}$$

$$r_2 = \frac{DH - \sqrt{D^2H^2 + 4\ i\omega D}}{2D} \tag{4.9}$$

We obtain a solution by adding the two solutions for $(+\omega)$ and $(-\omega)$ and going to the limit $\omega = 0$.

We notice that for small ω,

$$r_1 = H + i\,\frac{\omega}{DH}$$

$$r_2 = -i\,\frac{\omega}{DH}\ ; \tag{4.10}$$

we then obtain the solution

$$\langle x\rangle = \frac{h}{e^{Ha}-1}\left[\frac{\exp[H(a+z)]}{H} - (a+z-DHt)\right] \tag{4.11}$$

This corresponds to the superposition of a steady-state solution (reached in a very short time) and the solution expressing the slow filling of the convection zone with its newly formed elements. We can check if the boundary solutions are fulfilled, and then make the choice

$$h = \frac{\Phi}{2n_{H_o}\ D_o} \tag{4.12}$$

In the case of a deep convective zone, $e^{Ha} \gg 1$. Near the surface $(z \simeq 0)$, we can write

$$\langle x\rangle = h\left(\frac{e^{Ha}}{H} + DHt\ e^{-Ha}\right), \tag{4.13}$$

and we see that the major effect will be due to the balance of element formation and diffusion, with the cumulative effect of the transport into the convective zone, as in the following cases:

(a) for $DH^2te^{-Ha} < 1$, the surface elements are dominant;

(b) for $DH^2te^{-Ha} > 1$, the diluted elements are dominant.

The actual result is not entirely independent of the possible change of Φ and a, during stellar evolution. Let us call ξ, M, the mass of the convective zone.

If ρ_o is the density at level $z = 0$, we write

$$\frac{\rho_o e^{Ha}}{H} = \frac{\xi m}{4\pi R^2} \tag{4.14}$$

Therefore, we have to consider the inequality

$$t \lesseqgtr \frac{\xi m}{4\pi R^2} \quad \frac{1}{DH\rho_o} \tag{4.15}$$

The main problem to be discussed now is the value of D.

E. The Diffusion Coefficient.

First we consider the data from the hydrodynamic motion in the solar atmosphere.

If N is the frequency of the appearance of the granules, and L their characteristic size, then the horizontal diffusion coefficient is L^2N. According to Simon and Leighton (1964), we obtain the following values:

for the supergranulation, $D = 3.15 \times 10^{13}$ cm^{-1};

for the granulation, $D = 2.5 \times 10^{13}$ cm^{-1}.

We want to obtain the diffusion coefficient for the vertical motion.

We consider a velocity field given by:

$$\frac{dz}{dt} = V_z \sin (2\pi\rho/L)$$

$$\frac{d\rho}{dt} = V_\rho \cos (2\pi\rho/L), \tag{4.16}$$

in which ρ is the distance to the center of the granule. We can obtain the vertical displacement during horizontal displacement from ρ_1 to ρ_2:

$$z = \frac{V_z}{V_\rho} \quad \frac{L}{2\pi} \log \frac{\cos (2\pi\rho_1/L)}{\cos (2\pi\rho_2/L)} \tag{4.17}$$

By integration of the second equation of motion, one obtains the relation between the duration of the flow and the positions ρ_1 and ρ_2

$$\pi V_\rho t/L = \log \frac{\tan(\pi/4 + \pi\rho_2/L)}{\tan (\pi/4 + \pi\rho_1/L)} \tag{4.18}$$

In order to have an order of magnitude, we take the origin at the center of the granule, $\rho = 0$, and using the value $v_\rho/v_z = 4$, given by Simon and Leighton (1964), we obtain

$$z = 0.035\ L, \qquad (4.19)$$

from which we derive the vertical diffusion coefficient

$$D_z = 2.5 \times 10^{10}\ \ cm^{-1}. \qquad (4.20)$$

Now, we want to reinterpret the estimate of the vertical diffusion coefficient of other stars, especially of contracting and sub-giant stars. If we assume that the velocity of the turbulence in the convective zone is obtained by the classical phenomenological theory of the convective zone, we must proceed to calculate the buoyancy force. We have

$$F_{conv.} = C\ \frac{\rho}{2H}\ [\delta(d\log T/d\log P)]^{3/2}(g\mu/k)^{3/2}(T)^{1/2} \qquad (4.21)$$

If we take the convective flux at a reference level given by the optical depth, we can assume $(\mu\rho/H)$ to be constant. If, as an interpolation, we write $\mu{\sim}\rho T^{-n}$ for the absorption coefficient, we then obtain the homologous relation for the density

$$\rho \sim M^{1/2}\ T^{(n-1)/2}R^{-1}.$$

We then have

$$\left[\delta\left(\frac{d\ \log\ T}{d\ \log\ P}\right)\right]^{3/2} \sim R^3T^{(8-n)/2}M^{-3/2}$$

and

$$V \sim RT^{(11-n)/3}M^{-2}$$

from which we derive

$$V\ell \sim R^3T^{(17-n)/6}\ M^{-2}.$$

In order to get an idea of the homologous change of n, we take n = 5, and write,

$$V\ell \sim R^3T^2M^{-2}$$

We must consider the homologous transform of $DH\rho_o$, in equation ,
which leads to

$$DH\rho_o \sim v\rho_o \sim T^4M^{-3/2} \qquad (4.22)$$

For the characteristic time, we finally obtain:

$$t_c = \xi \times 1.6 \times 10^{15} \left(\frac{M}{M_\odot}\right)^{5/2} \left(\frac{T}{T_\odot}\right)^{-4} \left(\frac{R}{R_\odot}\right)^{-2},$$

$$= \xi \times 1.6 \times 10^{15} \left(\frac{M}{M_\odot}\right)^{5/2} \left(\frac{L_\odot}{L}\right) . \tag{4.23}$$

F. The Characteristic Time of Diffusion.

This characteristic time must now be compared with the time of gravitational contraction along the Hayashi track given by

$$t_H = c \frac{GM^2}{3RL} \tag{4.24}$$

with $T_{eff} \simeq$ constant, c = 3/7, and for a polytrope, n = 3/2. This gives

$$t_H = 1.4 \times 10^{14} \left(\frac{M}{M_\odot}\right)^2 \frac{R_\odot}{R} \frac{L_\odot}{L} . \tag{4.25}$$

The actual time at our disposal during the Hayashi phase will be smaller than the characteristic time t_c, as long as

$$\frac{R_\odot}{R} \left(\frac{M_\odot}{M}\right)^{1/2} < 11.4 , \tag{4.26}$$

which is certainly fulfilled, since $(R_\odot/R)<1$ during the Hayashi phase (1961) for stars of mass comparable to the solar mass.

If we use expression (4.25) for the time t_H, and expression (4.23) for the time t_c in equation (4.13), we obtain:

$$\langle x \rangle = \frac{\Phi}{2n_{H_o} D_o H} \left[e^{-1} + .087 \frac{R_\odot}{R} \left(\frac{M_\odot}{M}\right)^{1/2} \right]. \tag{4.27}$$

We conclude that, along the Hayashi track, the surface abundance is entirely determined by the ratio of flux of newly formed particles to the diffusion

$$\langle x\rangle \simeq \frac{1}{2e} \frac{\Phi}{n_{H_o} D_o H} \tag{4.28}$$

Knowing $\langle x\rangle$, we can calculate Φ. We have

$$\Phi \cong 2e\langle x\rangle \; (n_{H_o} D_o H)_\odot \left(\frac{T}{T_o}\right)^4 \left(\frac{M}{M_\odot}\right)^{-3/2} .$$

For $n_{H_o} = 10^{16}$, $H^{-1} = 3 \times 10^7$, $\langle x\rangle = 10^{-9}$, we obtain

$$\Phi \cong 4 \times 10^{19} .$$

For an energy of 10^{13}e.v., per Be formed, this gives a flux of,

$$\frac{L_{cR}}{4\pi R^2} \simeq 6 \times 10^{11} \left(\frac{T}{T_\odot}\right)^4 \left(\frac{M}{M_\odot}\right)^{-3/2} \text{ergs sec}^{-1}\text{cm}^{-1}$$

or

$$\frac{L_{cR}}{L_\odot} \simeq 10 \frac{L}{L_\odot} \left(\frac{M_\odot}{M}\right)^{3/2} . \tag{4.29}$$

This value is impossibly high, thus inducing a thermodynamic analysis of the accelerating mechanism. This will be done later.

The time of gravitational contraction along the Kelvin-Helmholtz scale is given by

$$t_{KH} = c \frac{GM^2}{RL} ,$$

with $c = \frac{3}{4}$; and, for a polytrope, n = 3. This gives

$$t_{KH} = 7.4 \;\; 10^{14} \left(\frac{M}{M_\odot}\right)^2 \frac{R_\odot}{R} \frac{L_\odot}{L} .$$

The actual time available during the Kelvin-Helmholtz phase will be smaller than the characteristic time, t_c, so long as

$$\frac{R_\odot}{R} \left(\frac{M_\odot}{M}\right)^{1/2} \frac{1}{\xi} < 2.2 . \tag{4.30}$$

As soon as the star ceases to be entirely convective, condition (4.30) is not fulfilled, and the abundance which is observed can be attributed to the progressive increases of concentration of Li and Be in the convective zone.

In this case we calculate:

$$\langle x \rangle = \frac{\Phi}{2 n_{H_o} D_o H} \quad 0.46 \, \frac{R_\odot}{R} \left(\frac{M_\odot}{M} \right)^{1/2} \frac{1}{\xi} ,$$

from which we derive Φ :

$$\Phi = \frac{2}{0.46} \, \langle x \rangle \, (n_{H_o} D_o H) \left(\frac{T}{T_\odot} \right)^4 \frac{M_\odot}{M} \left(\xi \frac{R}{R_\odot} \right) ,$$

with $n_{H_o} H \langle x \rangle \simeq 7 \; 10^{18}$;

this gives

$$\Phi = 3 \;\; 10^{10} \left(\frac{T}{T_o} \right)^4 \frac{M_o}{M} \frac{R}{R_o} \, \xi , \qquad (4.31)$$

or, for 16 ergs per Be formed, we must have

$$\frac{L_{oR}}{L_o} = 5\xi \, \frac{L}{L_o} \frac{M_o}{M} \frac{R}{R_o} . \qquad (4.32)$$

The situation is slightly better than that of an entirely convective star, because ξ is of the order of 0.1 or less; however, we must also consider this value as impossibly high.

G. Turbulent Diffusion. II.

The presence of a diffusion shield is difficult to accept. Reeves has suggested that the heating of the matter by the accelerated particles decreases the elastic cross section and follows the spallation process. This can be shown to be actually the case. However, this does not explain the remarkable constancy of the lithium abundance in young stars, which are not likely to have all the same activity. We are therefore led to the idea of an interstellar origin of the light elements. This is presently under verification by Reeves and co-workers (1970).

Then comes the problem of depletion. In order to explain the depletion of lithium, as shown by the ratio (Li/Be) by Merchant (1966) and by Conti and Danziger (1966), another mechanism has to be taken into consideration.

It is therefore appropriate to assume that the hydrogen convective zone rotates like a solid body and adjusts immediately to the angular momentum losses which take place at the surface of the Sun. As a consequence, the hydrogen convective zone will exert a torque on the interior of the Sun, and the angular momentum which is stored inside the Sun will flow to the surface. There has been some discussion about the mechanism of transport of angular momentum inside the Sun, as can be seen from the opposite views of Dicke (1970), and of Howard, Moore and Spiegel (1967). We shall take here the point of view, already expressed by Goldreich and Schubert (1967), and by Howard, Moore and Spiegel (1967), that the shear flow which is induced by the gradient of angular momentum inside the Sun generates a mild turbulence which is itself the agent of the transport of angular momentum. This leads to an estimate of the turbulent diffusion coefficient

$$D \simeq \frac{L^2}{T}$$

where L is the radius of the lower boundary of the hydrogen convective zone, and T the time scale of loss of angular momentum. With $L = 6.10^{10}$ cm, $T = 10^9$ years, this gives

$$D \simeq 10^4 \text{ cm}^2 \text{ sec}^{-1} \quad .$$

We are led to the idea that turbulent diffusion plays an important role in depletion of light elements. These three elements are not formed presently at the surface of the Sun. Their destruction by thermonuclear burning takes place at different depths.

The problem of these elements is the following. In very young stars, the surface abundance of lithium and beryllium is much larger than in the Sun [Wallerstein and Conti (1969)]. There is no doubt that depletion has taken place in the Sun. Until the 1967 discussion on the lithium problem [I.A.U. (1968)], it was currently considered that the destruction of the lithium takes place at the bottom of the hydrogen convective zone. However, as Böhm (1966) and Böhm and Stückl (1967) have shown, it seems impossible to bring the bottom of the convective zone sufficiently low for reaching the temperature of lithium burning. Furthermore, very simple considerations on the different rates of thermonuclear burning show that if the burning were taking place at the bottom of the convective zone, only one deficiency at a time can be explained. It is therefore necessary to seek another mechanism.

The following explanation is, in its principle, very simple. We shall assume that the bottom of the convective zone is much higher than the level of thermonuclear burning of lithium, beryllium and boron respectively. The time scale of burning will be entirely defined by the time scale of the transport by turbulent diffusion from the bottom of the hydrogen convective zone.

Our purpose here is to derive the value of the turbulent diffusion coefficient from the observations. For such a derivation, we need to know the distance of the bottom of the convective zone to the level of thermonuclear burning and we need to know also the amount of destruction which has taken place.

Concerning the level of thermonuclear burning, there is not much uncertainty. We shall define that level by the condition that the time scale of burning is much less than the age of the Sun, for example, one tenth of the age of the Sun. The greatest uncertainty is in the level of the lower boundary of the hydrogen convective zone. The distances ΔR from the bottom of the hydrogen convective zone to the level of thermonuclear burning which are used in the present paper can be found in Table 12.

For the estimate of the deficiency, we shall proceed in the following way. We assume that the initial abundances are in the ratio of the spallation cross-sections, as discussed, for example, by Bernas et al. (1968). We take for the initial value of the lithium abundance what seems to be a standard value for T Tau stars [Bonsack and Greenstein (1960), Bonsack (1961)] and for young stars [Wallerstein and Conti (1969)]. This defines the initial abundances of Table 12.

The determination of the present abundance is made in the

following way. For ^{7}Li , we take what seems to be a reliable value, due to Grevesse (1968) and to Peach (1968), log [^{7}Li] = 0.4 (for log [^{1}H] = 12). The abundance of ^{6}Li is much smaller. It certainly does not exceed the terrestrial abundance. The abundance of ^{9}Be has been determined by Hauge and Engvold (1968), who give log [^{9}Be] = 0.96 and by Grevesse (1969) who gives log [^{9}Be] = 1. However, a recent study of the limb darkening of the lines of BeII was led Muller (1970) to demonstrate the existence of a blend. Therefore, the value of Grevesse is an overestimate. It is not until a precise study will have been carried out that we shall know exactly the present abundance of ^{9}Be in the Sun. On a provisional basis we shall take log [^{9}Be] = 0.5. This is the value which is given in Table 12.

Boron has not been observed, and only an upper limit, log [B] < 2.7 can be given.

The ratio of the final to the initial abundances can be written

$$\frac{X_{final}}{X_{initial}} = \exp(-\xi_X) \quad ,$$

where X is any of the four elements or isotopes involved. A sufficient approximation consists in studying the diffusion in a plane parallel layer of thickness L = ΔR. This leads to the value

$$\xi_X = \frac{\pi^2}{4} \frac{D}{L_X^2} t$$

where t is naturally the age of the Sun. The method consists in writing

$$D = \frac{4}{\pi^2} \frac{L_X^2}{t} \xi_X$$

The values which have been obtained by that method are collected in Table 12.

Element	log initial abundance	log present abundance	ξ	ΔR ($\times 10^{10}$ cm)	D ($cm^2\ sec^{-1}$)
^{6}Li	2.7	<-0.7	>7.8	1.4	> 4 . 10^3
^{7}Li	3	0.4	6	1.9	5.5 . 10^3
Be	1.5	0.5	2.3	2.7	4.4 . 10^3
B	3.3	< 2.7	>1.4	3.8	>5.2 . 10^3

Table XII

Turbulent diffusion coefficient in the Sun from light elements burning

The remarkable agreement between the different determinations of the turbulent diffusion coefficient suggests very strongly that the turbulent mechanism which has been described here exists at least down through one-half of the solar radius.

BIBLIOGRAPHY

Chapter IV

AUDOUZE, J., GRADSZTAJN, E., and REEVES, H., (1967), "The Formation of Lithium in Young Stars," in "Instabilité gravitationnelle et formation des étoiles, des galaxies et de leurs structures caractéristiques," Mém. Soc. Roy. Sci. de Liège, XV, 299.

BERNAS, R., GRADSZTAJN, E., REEVES, H., YIOU, F., AND SCHATZMAN, E., (1967), "The Nucleosynthesis of Lithium, Beryllium, and Boron, and the Formation of the Solar System," Annals of Physics, 44, 426.

BÖHM, K. H., (1966), Z. Naturforsch., 21a, 1107.

BÖHM, K. H., AND STÜCKL, E., (1967), Z. Astrophys., 66, 487.

BONSACK, W. K., (1959), "The Abundance of Lithium and Convective Mixing in Stars of Type K*," Ap. J., 130, 843.

BONSACK, W. K., and GREENSTEIN, J. L., (1960), Ap. J., 131, 83.

BONSACK, W. K., (1961), Ap. J., 133, 551.

DICKE, R. H., (1970), Ap. J., 159, 1.

GOLDREICH, P., AND SCHUBERT, G., (1967), Science, 156, 1101.

GREVESSE, N., (1968), Solar Physics, 5, 159.

GREVESSE, N., (1970), Private communication.

HAUGE, O., and ENGVOLD, O., (1968), Astrophys. Letters, 2, 235.

HAYASHI, C., (1961), "Stellar Evolution in Early Phases of Gravitational Contraction," Publ. Astr. Soc. Japan, 13, 450.

HERBIG, G. H., (1965), "Lithium Abundancies in F5-G8 Dwarfs," Ap. J., 141, 588.

HERBIG, G., WALLERSTEIN, G., and CONTI, P. S., (1965), "Observations of the Lithium Content of Main Sequence Stars in the Hyades," Ap. J., 141, 610.

HOWARD, L. N., MOORE, D. W., and SPIEGEL, E. A., (1967), Nature, 214, 1297.

I.A.U., (1967), Proceedings of the XIIIth General Assembly, Praha.

MERCHANT, A. E., BODENHEIMER, P., and WALLERSTEIN, G., (1966), "The Lithium Isotope Ratio in Two Hyades F. Stars," Ap. J., 142, 790.

MERCHANT, A. E., (1966), "Beryllium in F- and G-type Dwarfs," Ap. J., 143, 336.

MORSE, M., AND FESHBACH, H., (1953), "Methods of Theoretical Physics," New York, McGraw Hill.

REEVES, H., (1966), "Nuclear Energy Generation in Stars, and Some Aspects of Nucleosynthesis," Stellar Evolution, edited by R. Stein and A. Cameron, Plenum Press, New York, 83.

SIMON, G. W., AND LEIGHTON, R. B., (1964), "Velocity Field in the Solar Atmosphere, III. Large-Scale Motion, the Chromospheric Network, and Magnetic Fields," Ap. J., 140, 1120.

SYROVATSKII, S. I., (1961), "Spectrum of Galactic and Solar Cosmic Rays," J.E.T.P., 13, 1257.

TSYTOVITCH, V. N., (1963), "Acceleration by Radiation and the Generation of Fast Particles in Cosmic Space," Astron. J. U.S.S.R., 40, 612.

WALLERSTEIN, G., and CONTI, P. S., (1969), Annual Rev. Astr. and Astrophys., 7, 99.

V. THE ROLE OF ELECTROMAGNETIC ACTIVITY IN STELLAR EVOLUTION

A. Introduction.

The present discussion is an attempt to connect a certain number of facts, with the help of only a few new hypotheses. It is based essentially on a report of the author (Schatzman, 1960). Since the starting point of this work is a discussion of the difference in the speed of rotation of stars on the upper and lower parts of the main sequences, it is worthwhile to understand clearly the logical implication of that well-known fact.

The fact that early type stars have a larger equatorial velocity implies that they have been formed by contraction. Indeed, a contracting star always has a tendency to increase its angular velocity, unless some powerful mechanism acts as a brake. We can therefore rule out the suggestion that stars and interstellar matter have been formed by expansion of the same prestellar matter, since we cannot imagine any way of storing a large amount of angular momentum in a nonrotating fluid.

Mestel and Spitzer (1956) have shown the kinds of difficulties met when considering the process of contraction of a sphere of gas. We shall assume, in the following, that these difficulties have been solved, e.g., that the magnetic field can decay during contraction, and that the cloud is able to break up into stars.

If we start with a sphere of low-density interstellar gas, we can estimate the density it must have to reach the main sequence with the maximum possible angular velocity, if the initial angular velocity of the gas is the angular velocity of the sun around the galactic center. If R_o is the radius of the sphere of interstellar gas, ω_o its angular velocity, R the radius of the star, ω its angular velocity, $(2/5)MR_o^2$ the moment of inertia of the cloud, KMR^2 the moment of inertia of the star, we have:

$$KMR^2\omega = \frac{2}{5} MR_o^2\omega_o \; . \tag{5.1}$$

As we suppose that the mass has not changed, we have:

$$\frac{R}{R_o} = \left(\frac{2}{5K}\,\frac{\omega_o}{\omega}\right)^{1/2} . \qquad (5.2)$$

Let us take for ω_o the angular velocity of the galaxy, which is of the order of 10^{-15} sec^{-1}. Suppose also that ω is the maximum angular velocity ω_{max}. Writing the equality between the gravity and the centrifugal force at the equator, we obtain the approximate value $\omega^2_{max} = \frac{4\pi}{3} G\rho$. For a star like the sun, $\omega_{max_o} \simeq 10^{-3.45}$.

Finally, for the standard model, $1/K = 13.5$: we shall always take the same value for the coefficient of the moment of inertia.

With these values, we obtain:

$$\frac{\rho_o}{\bar{\rho}} = 10^{-16} .$$

With $\bar{\rho} = 2$ g cm^{-3}, this leads to $\rho_o \simeq 2.10^{-16}$ g cm^{-3}, which is a very high density. A similar conclusion was reached by Spitzer (1948). If we start with densities of the order of 10^{-23}, of the same order of magnitude as the density of interstellar matter, and with the same angular velocity $\omega_o \simeq 10^{-15}$ sec^{-1}, the contracting proto-star will become flat and lose matter through the equator as soon as it reaches the density given by the relation:

$$\bar{\rho}_i = \left(\frac{4\pi}{3}\right)^3 \rho_o^4\, \omega_o^{-6} \qquad (5.3)$$

For $\rho_o = 3.10^{-24}$, $\omega_o = 10^{-15}$, $\bar{\rho}_i = 7.4\ 10^{-3}$: This corresponds to the density of a giant star. Further contraction would be accompanied by equatorial mass loss. Therefore, we should expect all stars along the main sequence to have a very fast rotation. Such a conclusion has been known for a long time; its first mention can probably be found in the criticism by Fouche and by Babinet of Laplace's cosmogony (see: Schatzman, [1967]).

A homology transformation will help us to find the track of a star during its contracting phase. If we suppose that the star has always the same internal structure, we have two relations.

One expresses the fact that, at the equator, gravitational and centrifugal force are equal:

$$\lambda_1 \frac{GM}{R^2} = \omega^2 R \tag{5.4}$$

in which λ_1 is a constant, of order of magnitude unity, which takes into account the flattening of the system. The other equations, which express the conservation of angular momentum, suppose that angular momentum is lost only at the equator, with an angular momentum $R^2\omega$ per gram:

$$\frac{d}{dt} (KMR^2 co) = R^2 \omega \frac{dM}{dt} \tag{5.5}$$

Dividing both sides by $KMR^2\omega$ we obtain

$$1 - \frac{1}{K} \frac{1}{M} \frac{dM}{dt} + \frac{2}{R} \frac{dR}{dt} + \frac{1}{\omega} \frac{d\omega}{dt} = 0 \ . \tag{5.6}$$

Integrating equation (5.6) we obtain the homology relation

$$M^{1-1/K} R^2 \omega = \text{constant}, \tag{5.7}$$

If we insert in equation (5.7) the relation (5.4), we obtain the homology relations

$$R \sim M^{2/K -3}$$

$$\omega \sim M^{-3/K +5} \tag{5.8}$$

As we have seen, $1/K$ is of the order of 13.5. Consequently, during that phase, the mass loss is small since

$$M \sim R^{1/24} \tag{5.9}$$

The angular velocity increases slowly, as

$$\omega \sim R^{-71/48} \tag{5.10}$$

and there is an even slower increase of equatorial velocity,

$$V = \omega R \sim R^{-23/48} \tag{5.11}$$

To place this homologous model in the Hertzsprung-Russell diagram, we consider the homology relation for the main sequence, and the homology relation for the contracting star without mass loss. Converting the mass luminosity relation into a (L, T_{eff}) relation, we obtain:

main sequence	$L \sim T_{eff}^{6.4}$
contracting star (constant mass)	$L \sim T_{eff}^{0.363}$
contracting star (equatorial mass loss) according to equation	$L \sim T_{eff}^{0.0615}$

One can find(Fig. 22)the path of a star in the Hertzsprung-Russell diagram, where it has been assumed that the equatorial mass loss begins some time before reaching the main sequence.

The result mentioned here is slightly different from the result formerly published (Schatzman, 1949), due to a different expression of the opacity law, implicitly contained in the homology relations.

Let us now consider the various suggestions which have been made to explain the loss of angular momentum during star formation.

It has been suggested that stars lose mass from their surface by corpuscular radiation; Fessenkoff (1949) and Schatzman (1954 a,b) have considered the possibility that such a phenomenon could explain the difference between the early and late-type stars. However, three major criticisms can be made of that theory: (1) To slow down the motion of a star and make it evolve from type A to type F, it is necessary for it to lose about half of its mass. Such a large mass loss would hardly escape observation. (2) The theory does not explain the sharp difference between stars earlier and later than type F2. (3) The distribution among giants and subgiants, well explained by conservation of

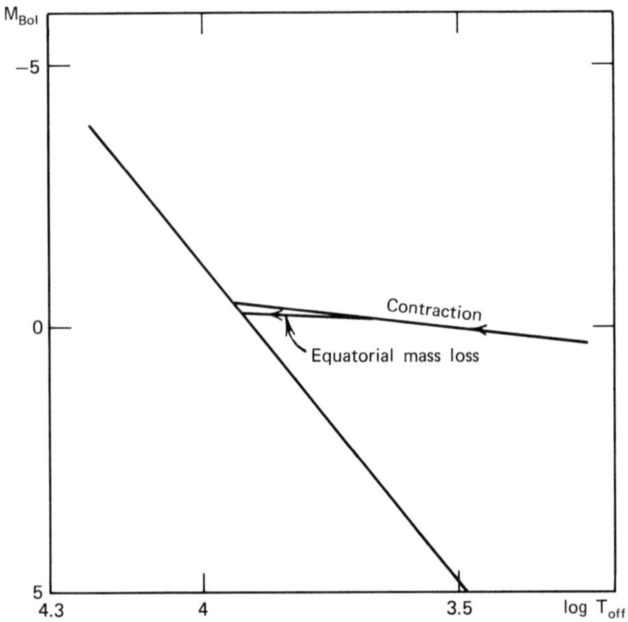

FIG. 22—Diagrams of Hertzsprung-Russell (M_{Hob} log T_{eff}) showing the main sequence, the homologous contraction without and with equatorial mass loss.

angular momentum (Oke and Greenstein, 1954), is left completely out of consideration by the theory of corpuscular radiation.

The theory that stars could lose angular momentum through magnetic interaction with the interstellar medium has had some attention (Spitzer, 1956), the sharp difference at type F2 possibly being due to the difference in size and density of the Strömgren spheres. However, if we consider the distance at which matter can be dragged with the rotating magnetic field, we find that it is inversely proportional to the square root of the density of the circumstellar matter; therefore, the rate of transfer of the angular momentum from the rotating interstellar matter to the matter at rest, through any kind of friction, is larger in an H II region than in an H I region.

Lüst and Schlüter (1955) have considered the loss of angular momentum from magnetic coupling between a star, its magnetic field and an envelope of conductive matter. The rate of loss of angular momentum is essentially related to the decay of the magnetic field and is a very slow and inefficient process (see also L. Mestel, 1959).

B. Electromagnetic Coupling.

Another mechanism, which combines coupling with the stellar magnetic field and mass loss has already been proposed by Schatzman (1959).

The idea is the following: we suppose that when ejection of matter occurs at the surface of a star which has a magnetic field, the ejected matter is forced to turn, with the star, up to a critical distance beyond which it is no longer dragged by the magnetic field. Since that distance is much larger than the radius of the star, the loss of angular momentum per unit mass is large; a small mass loss can produce a very large effect.

More precisely, we could describe the mechanism in the following way: during flares, jets of matter are produced which, after having locally disturbed the magnetic field, follow the magnetic lines of force up to a point where the magnetic stresses are no longer sufficiently large to force matter to follow the lines of force. At that point, it can be said that matter leaves the star and carries away some angular momentum.

In order to estimate the rate of loss of angular momentum, we follow an idea of Cowling. If B is the magnetic field at a distance r from the star, we suppose that the field drags the matter, as long as the Alfvén velocity backward to the star is larger than the material velocity. The distance at which the loss of angular momentum takes place is difficult to estimate, as it seems that the matter is ejected from the sun at about Alfvén velocity.

However, the picture is a little more complicated, as we have to consider that the ejected matter is guided by lines of force, and that we have to compare the velocity of matter inside the tube of the magnetic field to the Alfvén velocity in the matter of the guide.

If we judge from the guiding of low-energy solar cosmic rays, guided by the magnetic field of the streams to the distance of the earth, we can conclude that guiding is a very efficient process.

Similar results for the solar wind were obtained by Brandt (1966).

Let us now consider the equation of conservation of angular momentum. We suppose that the loss of momentum occurs at a distance r larger than the radius R of the star, and then the equation of conservation of angular momentum (5.5) becomes:

$$\frac{d}{dt}\left(KMR^2\omega\right) = \frac{r^2}{R^2\omega}\frac{dM}{dt} \tag{5.12}$$

If $\frac{r}{R}$ is constant, we obtain:

$$\left(\frac{\omega}{\omega_1}\right)^{2/5} = 1 + \frac{2}{5}\left(\frac{r}{R}\right)^2 \frac{\Delta M}{m} \tag{5.13}$$

if we suppose that M changes very little. Let us take the largest possible value for ω_1, corresponding to the equality between the gravitational and the centrifugal force at the equator:

$$\omega_1^2 = \omega_{max}^2 = \frac{GM}{R^2} \tag{5.14}$$

the numerical value of ω_{max} is $10^{-3.45}$ for the solar mass and radius.

With $(r/R) \simeq 10^2$, we obtain

$$\frac{\omega}{\omega_1}^{2/5} = 1 - 4.10^{-3} \frac{\Delta M}{M} \tag{5.15}$$

in which ΔM is the mass loss.

We thus see that a very small mass loss can lead to an enormous loss of angular momentum, and that it would even be possible that the star be brought to rest.

C. The Convective Zone.

The physical process we have been considering does not, in itself, lead to an explanation of the difference between early and late type stars. However, the process implies both an ejection of matter and a magnetic field. We shall suppose that the ejection of matter is due to the surface activity, similar to the activity at the surface of the sun.

Surface activity is related to a cycle of magnetic activity in the sun. Ejection of matter is due to flares, which occur in the regions of complex magnetic spots, as has been proven by the observations of Severny (1958), Howard (1959), and Leighton (1959). The neutral-line theory of solar flares (Sweet, 1958; Dungey, 1953) does not seem to be confirmed by the observations. If we accept Parker's picture (1955) of solar activity, we are bound to establish a correlation between the presence of the surface convective zone (hydrogen convective zone) and the surface activity.

We shall therefore assume that whenever a convection zone exists, surface activity is a normal feature of the star, but that when the convective zone is absent or small, the surface activity is also small or absent even though a strong magnetic field might be present.

As is well known, G, K and M stars have a convective zone of considerable thickness (hydrogen convective zone) and A and B have a thin convective zone or none, except that probably the early B stars have a helium convective zone.

The transition from a thick convective zone to a thin convective zone is quite sharp. If we consider the two extreme cases studied by Ch. Pecker-Wimel (1953) and E. Bohm-Vitense (1951, 1953), we can conclude that the actual case is intermediary. The transition probably takes place in the F-type stars.

This has been confirmed by Baker and Kippenhahn (1963). The thickness of the convective zone has been calculated from main sequence stars, for a mixing length equal to 1 or 2 scale heights of the pressure. It is remarkable to notice (Fig. 5) how quickly the convective zone deepens, near to spectral type F2. This can be extrapolated to other stars.

For stars of a given chemical composition, the line which divides the Hertzsprung-Russell diagram into two regions, region C where stars have a convective zone, and D where stars have a very shallow convective zone or none, is almost parallel to the vertical axis (Fig. 23).

Starting in the F types on the main sequence, it reaches the G types in the supergiant range.

It is supposed that when a contracting star goes from region C to region D (Fig. 23), it stops losing angular momentum and therefore can reach the main sequence with a high equatorial velocity. The helium convective zone (region C', Fig. 23) can slow down the stars which happen to reach the main sequence in its range. However, due to its small extent, region C' does not have the power to completely stop the stars which cross it. Therefore, we can only expect a small effect.

We therefore have found what seems to be the basic explanation of the distribution of equatorial velocities among the main sequence: the connection between the convective zone, the magnetic activity (surface activity) and the loss of the angular momentum.

D. Energy Balance.

The energy needed for the exchange of angular momentum between the star and the ejected material is borrowed from the magnetic field. The total energy of a system of magnetic spots is of the order of a, where a is the distance from one spot to the other, and B_0 the magnetic field in the spots. This has to be compared to the energy loss during flares:

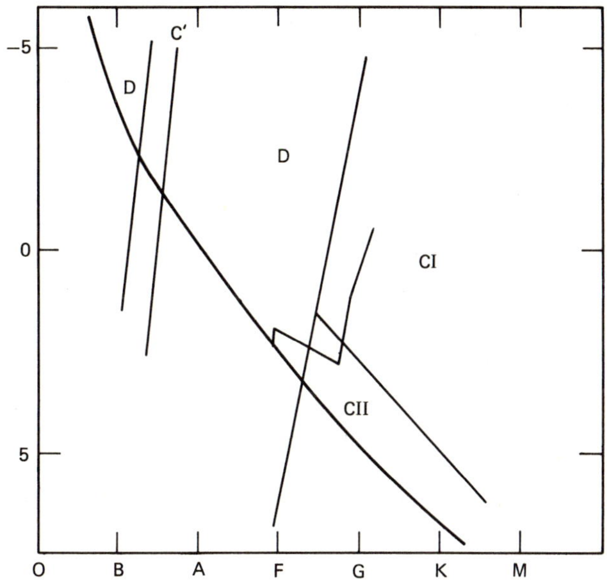

FIG. 23–Regions in the H–R diagram. Regions *C* and *C′* correspond to stars with a convection zone, *D* without a convective zone. In region *C*I, stars present electromagnetic and equatorial mass loss; in region *C*II, only electromagnetic mass loss.

$$\frac{1}{2} V^2 \delta M < \frac{B_o^2 a^3}{8\pi} \qquad (5.16)$$

With the values we used earlier ($V = 10^8$, $a/R - 0.1$, $B_0 = 10^4$), we find that a group of magnetic spots has stored enough energy to allow a maximum mass loss $\delta M < 10^{20}$ g. Let us consider a total mass loss of 10^{30}g. This represents a series of 10^9 groups of flares, each group being produced in the same magnetic region. If that succession is to occur in 10^7 years, it requires 10^2 groups of flares a year. Such a frequency is not improbable, from what we know, for example, as in UV Ceti and T Tauri stars. If a very small part of the magnetic energy is used to produce the ejection of matter, we can consider the possibility that the number of magnetic spots is larger, or that flares will start from many different places at the surface of the star. Even an efficiency of 10^{-2} would be sufficient to provide the necessary mass loss.

If we now consider the total magnetic energy at the surface of a magnetic star which, according to the value of a magnetic field, is of the order of 10^{40} erg, we see that the energy stored in the magnetic field is quite sufficient to provide the energy necessary for the required mass loss.

E. Other Observational Evidence.

Wilson(1966) has suggested using the intensity of the emission in the K line as an index of activity. The picture is, then, that it is possible to relate the index of activity to the velocity of rotation.

The deceleration depends on the strength of the magnetic field, the initial rotational velocity, the velocity and the density of the wind, and the length of time over which the process acts.

The last factor is controlled by the time a star spends in regions of the Hertzsprung-Russell diagram where subsurface hydrogen convective zones give rise to concomitant chromospheric activity.

Wilson (1967) finds that as one passes down the main sequences, slow rotation sets in very abruptly (near F5V) at a value of B-V less than 0.02 mag. away from the place where K-line emission first puts in an appearance. Wilson advances the view that chromospheric activity and thus K-line intensity decay with advancing age.

Since, at a given spectral type, the stars of the youngest cluster studied by Wilson, Viz., the Pleiades, have the strongest emission lines, and the older, Hyades, Coma and Praesepe stars, have weaker emission, one can roughly express the strength of emission as a function of the nuclear time scale of the clusters.

Kraft (1967) then shows that solar-type stars are rotationally decelerated, even after reaching the main sequence, and that this deceleration takes place on a time scale considerably shorter than 10^9 years.

Table XIII

Rotational Velocity as a function of age for stars of mass $M/M_o = 1.2$ from Kraft (1967)

Type of Star	Approximate Nuclear Age (years)	<v> km/sec
Field non emission	5×10^8 to 10^9	6
Field emission	similar to Hyades	25
Hyades	4×10^8	18.5
Pleiades	3×10^7	39

F. Differences between Am and Ap Stars.

It is suggested that in Ap stars magnetic field stops convection, whereas in Am stars the tendency to build an inversion of density generates a Rayleigh-Taylor instability which might produce the micro-turbulence observed in Am stars.

G. Michaud (1969) has suggested adding to the forces acting on test particles the radiation pressure.

Let us sketch the theory.

In the case of absorption, the force exerted by a radiation flux φ_ν $d\nu$ on an atom will be

$$F_\lambda \, d\lambda = \frac{N_n}{N_{tot}} \sigma_n \varphi_1 \frac{d^\lambda}{\lambda^2}$$

For hydrogenoid atoms, and photoionization

$$\sigma_n^{(c)} = 7.9. \frac{n}{Z_n^2} \frac{1}{Y^3} 10^{-18} \text{ cm}^2$$

with $y = h\nu/\chi$. In fact, all the impulse does not get into a motion of the ion, and the fraction of the momentum given to the ion is

$$1 - \frac{8}{5} K = 1 - \frac{8}{5} (1 - \frac{1}{y})$$

from which we write

$$F_{ion} = \frac{N_n}{N_{tot}} \int (1 - \frac{8}{5} k) \sigma_n \frac{\varphi_\nu d_\nu}{c}$$

For φ_ν we take, in the case of great optical depth

$$\varphi_\nu d_\nu = \frac{4\pi}{3} \frac{\partial B_\nu}{\partial T} \frac{-1}{\varkappa_\nu \rho} \frac{\partial T}{\partial r} d\nu .$$

For a line, we have

$$F_{line} = \frac{N_n}{N_{tot}} \int \sigma_n \frac{\varphi_1 dr}{c}$$

For a strong line, $k_\nu \rho = k_{cont} \rho + N_n \sigma_n$ (N_n per unit volume), and using

$$\varphi_\nu d_\nu = \frac{4\pi}{3} \frac{\partial B_\nu}{\partial T_\nu} = - \frac{4\pi}{3} \frac{\partial B_\nu}{\partial T} \frac{1}{\varkappa_\nu \rho} \frac{\partial T}{\partial r}$$

it is possible to calculate the force.

It should be noticed that the force depends on the concentration of the supported atoms, as k_ν depends on that concentration. If the atom can be supported at all, its concentration reaches a maximum. For a higher concentration, the radiation pressure cannot support the atom and it falls inside the star; for a lower concentration, it is pushed upwards.

For a faint line, the apparent acceleration per atom is of the order of

$$g_{rad} = \frac{1}{m_A} \frac{\pi e^2}{m_e c} f \frac{\partial B_\nu}{\partial T} \frac{\partial T}{\partial r} \frac{1}{\varkappa_\nu \rho c}$$

This can easily be larger than the gravity. Even taking into account the fact that the thermal diffusion may be large, this leads possibly to a large value of the force.

Inside a line, when the line becomes strong, a saturation effect takes place. The integration has to be carried on the quantity

$$\frac{\pi e^2}{m_e c} f\, g(\nu) d\Delta\nu \frac{1}{\varkappa_{cont} + \frac{N}{N_{cont}} \frac{\pi e^2}{m_e c} f\, g(\nu)} .$$

For a Doppler width

$$g(\nu) d\Delta\nu = \frac{1}{\sqrt{\pi}} \frac{d\Delta\nu}{(\Delta\nu)_D} \exp \left(- \left(\frac{\Delta\nu}{(\Delta\nu)_D} \right)^2 \right)$$

which gives approximately

$$g_{rad} = - \frac{1}{m_A} \frac{1}{N_{at}} \frac{\Delta\nu_D}{c} \varkappa \frac{\partial B_\nu}{\partial T} \frac{\partial T}{\partial r}$$

with

$$x^2 = \log \left(\frac{N_{at}}{N_{cont}} \frac{1}{\sigma_{cont}} \frac{\pi e^2}{m_e c} \frac{f}{(\Delta\nu)_D} \right)$$

writing that $g_{rad} < g$ gives an upper limit to the value of N_{at} .

As Michaud (1969) has shown, this can lead to an enrichment of the surface layers of stars by some elements (case of the peculiar A stars).

The interesting problem is to find out what takes place at the bottom of the convective zone and to find the difference between a region with a convective zone and a region where the magnetic field has stopped convection.

BIBLIOGRAPHY

Chapter V

BABCOCK, H. W., (1958), "A Catalog of Magnetic Stars," _Ap. J. Suppl._ 30, 141.

BAKER, N., and KIPPENHAHN, (1963), Tables of Models of Stellar Convective Zones, Institute for Space Studies; New York.

BRANDT, J. C., (1966), "Consequences of Torque Exerted on the Sun by the Solar Wind," _Ap. J._, 144, 1221.

DICKE, R., (1958), "Gravitational Theory and Observation," _Physics Today_, 20, 55.

DUNGEY, J. W., (1953), "Conditions for the Occurrence of Electrical Discharges in Astrophysical Systems," _Phil. Mag._, 44, 725.

FESSENKOFF, V. G., (1949), "On the Problem of Cosmogony in Contemporary Astronomy," _A. J. U.S.S.R._, 26, 27.

HOYLE, F., (1960), "On the Origin of the Solar Nubula," _Quarterly Journal of the R. A. S._, 1, 28.

HOWARD, R., (1959), "Observations of Solar Magnetic Fields," _Ap. J._, 130, 193.

KRAFT, R., (1967), _Ap. J._

LEIGHTON, R. B., (1959), "Observations of Solar Magnetic Fields in Plage Regions," Ap. J., 130, 366.

LÜST, R., and SCHLÜTER, A., (1955), "Drehimpulstransport durch Magnetfelder und die Abbremsung rotierender Sterne," _Zs. f. Ap._, 38, 190.

MESTEL, L., and SPITZER, L., JR., (1956), Star Formation in Magnetic Dust Clouds, _M. N._, 503.

MESTEL, L., (1959), "A Note on the Magnetic Braking of a Rotating Star," _Mon. Not. R.A.S._, 119, 249.

MICHAUD, G., (1970), _Ap. J._ 160, 641.

OKE, J. B., and GREENSTEIN, J. L., (1954), "The Rotational Velocities of A-, F-, and G-type Giant Stars," _Ap. J._, 120, 384.

PARKER, E. N., (1955), "Hydromagnetic Dynamo Models," _Ap. J._, 122, 293.

PECKER-WIMEL, Ch., (1953), "Contribution à l'étude de la zone convective des étoiles," Ann. d'Ap., 16, 321.

SCHATZMAN, E., (1949), "On certain paths of stellar evolution. I. Preliminary Remarks," Acad. Ray. des Sciences de Belgique, 35, 1141.

SCHATZMAN, E., (1954a), "Remarques sur l'évolution des étoiles. III. Ann. d'Ap., 17, 300.

SCHATZMAN, E., (1954b), "Remarques sur l'évolution des étoiles. V., Ann. d'Ap., 17, 383.

SCHATZMAN, E., (1959), "Sur la perte de masse et des processus de freinage de la rotation," I.A.U. Symposium No. 10, edited by J. Greenstein, Paris, p. 129.

SCHATZMAN, E., (1966), The Origin and Evolution of the Universe, New York, Basic Books.

SEVERNY, A. B., (1958), News of the Crimean Observatory, 20, 22.

SPIEGEL, E., (1966), "Private Communication."

SPITZER, L., (1948), "The Formation of Cosmic Clouds," Centennial Symposia December 1946, Harvard Observatory Monographs No. 7, Published by the Observatory, Cambridge, Mass., p. 87.

SWEET, P. A., (1958), "The Neutral Point Theory of Solar Flares," I.A.U. Symposium No. 6, Electromagnetic Phenomena in Cosmical Physics, p. 123.

VITENSE, E., (1951), "Über die mittleren Zustandsgrössen und spektralen Eigenschaften von Sternatmosphären in Abhängigkeit von Effektivtemperatur und Schwerebeschleunigung," Zs. f. Ap., 29, 73.

VITENSE, E., (1953), "Die Wasserstoffkonvektionszone der Sonne," Zs. f. Ap., 32, 135.

WILSON, O., (1966), "Stellar Convective Zones, Chromospheres and Rotation," Zs. f. Ap., 32, 135.

WILSON, O., (1967), Ap. J.

DYNAMICS OF INTERPLANETARY MATERIAL

by

L. Biermann

Director, Max Planck Institute for Astrophysics

Munich, Germany

Lectures Presented During the Summer Session, 1968

in the

DEPARTMENT OF AEROSPACE ENGINEERING SCIENCES

University of Colorado

Boulder, Colorado

I. PHYSICS OF THE SOLAR CORONA

General Properties of the Interplanetary Plasmas

A. Introductory Remarks. The interplanetary materials--the particles and the fields embedded in the interplanetary material--have their origin predominantly in the sun; the solar corona is the source of the quiescent interplanetary gas ("solar wind"), which pervades with hypersonic velocity all interplanetary space. Since in this course of lectures the main emphasis will be given to the dynamical phenomena in interplanetary space, we shall report only on those observational facts related to the corona and discuss only those theoretical models of it which are of immediate importance to our subject. That is to say, we disregard or mention only in passing the relations between coronal hydrodynamics and the lower solar atmosphere (chromosphere and hydrogen convection zone); for recent account of these subjects I would like to refer to the monographs of D. Billings[1)] and of H. Zirin[2)], to the review article of G. Newkirk[3)], and to the Proceedings of the 5th Symposium on Cosmical Gas Dynamics, edited by R. N. Thomas[4)].

One may ask where the outer boundary of interplanetary space is located. As will be discussed in the last lecture, the region pervaded by a hypersonic plasma flow is probably at least as extended as that in which the larger planets move. The transition to subsonic conditions should take place at distances not very much larger than the orbit of the outermost known planet, Pluto. Really interstellar conditions are reached, however, only at much larger distances (see section VI).

The title of this course is essentially the same as that of the monograph[5)] published five years ago by E. Parker, to whom the discovery of the relationship between interplanetary and coronal hydrodynamics is due. From a number of more recent review articles I shall make use particularly of those by W. Axford[6], A. Dessler[20], R. Lüst[7], and E. Parker[8]. Further work, both observational and theoretical, which has greatly increased our understanding of interplanetary phenomena, will be quoted upon occasion.

Owing to the large mean free path of all particles in interplanetary space (including the outer corona) and to the presence everywhere of magnetic fields, it is important to realize the limitations of a purely hydrodynamical approach, notwithstanding the fact that this has been remarkably successful in elucidating the large scale phenomena. Under such circumstances the presence of the heat flow, which carries most of the energy out to 20 or 30 solar radii, has the important consequence that the energy equation cannot be formulated with the same degree of accuracy as those expressing the balance of mass and of momentum. For a description of the underlying pertinent physics, we shall make use particularly of Spitzer's monograph[9], which is almost exclusively concerned with the equations of continuity and of momentum, together with Maxwell's equations. For some problems, for instance of cometary plasmas, the changes in the macroscopic equation due to the addition to the plasma of new ions are important. For other problems of interplanetary physics it will even become important to use levels of description which are in a sense deeper than those of ordinary gas dynamics. The fact that in a number of ways the limits of the applicability of the usual macroscopic formulations are reached, may perhaps be regarded as a point of general interest of interplanetary plasma dynamics.

To the astrophysicist, the importance of studies of the solar corona, of the plasma tails of comets, and of interplanetary space is based on the fact that these are the only dilute cosmic plasmas observable at close range and that the interplanetary plasma is the only cosmic plasma in which we can make measurements and even experiments in situ, for instance on the transmission of radio signals or with artificially created plasma clouds which, like the comet tails, are observable from the ground against the night sky by ordinary optical means.

The general plan of these lectures is as follows: An outline of some relevant properties of the solar corona is followed by a description of the macroscopic properties of the solar wind. The hydrodynamical theory of the outer corona and the solar wind is first discussed on the basis of the Bernoulli equation, with only gross assumptions on the relation between pressure and density. Of the more refined attempts to arrive at realistic models of the solar wind by means of macroscopic equations including the energy equation, we discuss a scheme in which conduction and viscosity are included in the momentum and the energy equation, as well as one in which the different temperatures and conductivities of the ions and the electrons are explicitly taken into regard.

In order to introduce the subject of the limitations of macroscopic theories and of microscopic effects, some results of kinetic theory are used to get information on the general properties of the pressure tensor of ions and electrons. An outline of the derivation of macroscopic equations from the Boltzmann and the Maxwell equations, including the important limiting case of zero resistivity, is then given as a basis for a discussion of the observed distribution of the microscopic velocities, of small scale equilibria and discontinuities, and for the later sections on the magnetohydrodynamics of interplanetary space. Of this latter area, attention will be given to the transport of rotational momentum, to small scale features and the microscopic velocity distribution, and to the interaction of the solar wind with the interstellar gas surrounding the heliosphere, in which the gains and losses of mass, momentum and energy of the plasma are important.

B. Some Facts on the Physics of the Solar Corona. The transition from chromospheric (temp. $\approx 10^4$ °K) to coronal conditions ($T \approx 10^{6.2}$ °K) takes place over a distance of the order of or less than ≈ 1000 km, at an almost constant pressure of ≈ 0.1 dyn/cm^2 [10]. The ion number density at the level $1.0 \cdot 10^9$ cm ($= \frac{1}{70} R_\odot$) above the solar limb is some 10^8 cm^{-3} in the equatorial belt, the scale height $\approx 10^{9.9}$ cm. The temperature decreases outwards only very slowly, $d(\log T)/d(\log \rho)$ being ≈ 0.1; and the inner and middle corona outside of active regions appear to be in an approximately stationary state of pressure equilibrium, though the momentum flux of turbulence if present has to be taken into account. The high temperature which, in addition to a conductive loss downward[11] of $5 \cdot 10^5$ erg/cm^2 sec, leads to a radiative energy loss[1] of $\approx 1/2 \cdot 10^5$ erg/cm^2 sec, is thought to be maintained by progressive waves and oscillations, transferring non-thermal (mechanical and magnetic) energy from deeper layers up to the corona[12]. Recently observational evidence has been obtained by radar measurements[13] that at the level $r = 1.5\ R_\odot$ there always exist velocities of 150 km/sec --Mach number close to 1--which under the circumstances ($N_i \approx 2 \cdot 10^7$ cm^{-3}) can only be some type of wave or shock[14]. These velocities, which seem to slow down at higher layers, appear to confirm in a general way the correctness of the theoretical picture mentioned before. Another result from radar measurements is that at the same level a general outward motion of the order of 10 km/sec exists, which at the distance of the earth would correspond to a flux of the order of 10^9 ions/cm^2 sec or perhaps somewhat less if the flow is somewhat smaller in polar regions. This at least suggests a general outflow of the coronal gases in approximately the quantity indicated by the measurements from space vehicles ($\approx 10^{11.0}\ \frac{g}{\text{sec,ster}}$)

and by the theory to be discussed below. We note here only that even at the level $r = 2.5\ R_\odot$ the acceleration connected with this outward flow is still of the order of only 1% of the gravitational acceleration $g_\odot(10^{3.64}\ cm/sec^2$ at $2.5\ R_\odot)$; for $r \gtrapprox 4R_\odot$ it becomes $\gtrapprox 0.1\ g_\odot$, and for $r \gtrapprox 11\ R_\odot$ it is $\gtrapprox g_\odot$.

Direct evidence, from measurements in interplanetary space, on the state of ionization (and thereby on the temperature) of those parts of the corona from which the solar wind emanates, has been obtained recently by Hundhausen, Gilbert and Bame[15]. The measurements of the ratios O^{+5}: O^{+6}: O^{+7} indicate that the temperatures in the inner corona are usually ≈ 1.5 or $1.6 \cdot 10^6$, in perfect agreement with the older picture based on the spectrum and the theory of the ionization equilibrium; in some cases, in the presence of an active region, values $> 2.2 \cdot 10^6$ were found (O^{+7} being then more abundant than O^{+6}).

The Table I contains values of the electron density in the solar corona for equatorial regions near minimum solar activity (at maximum the figures are roughly twice as high) as given by Newkirk[3], together with some quantities derived from them. The total number densities are estimated under the assumption that there are 8% less ions than electrons; this corresponds to a mass content of He of $\gtrapprox 25\%$ and of 2 or 3% of heavier elements. The mean mass per particle is then very close to $10^{-24.0}$ g. The fourth column gives effectively $d(\log N)/d(\log r)$; the fifth column will be explained later. The values are representative averages; the variations with local and general activity of the sun are large (cf. 3).

We note that with a flux of $10^{35.0}$ ions + electrons per sec and ster ($4/9 \cdot 10^9\ cm^{-2}\ sec^{-1}$ at 1 a.u.), which

VALUES OF ELECTRON DENSITY IN THE SOLAR CORONA FOR EQUATORIAL REGIONS NEAR MINIMUM SOLAR ACTIVITY

Table I

$r/R_\odot$	N_e	$\log (N_e + N_i) \equiv \log N$	$\frac{\Delta \log N}{\Delta \log r}$	$\frac{r}{R_\odot} \cdot \frac{\Delta \log N}{\Delta \log r}$
(1.02)	$(400 \cdot 10^6)$			
1.2	$70 \cdot 10^6$	8.14	6.9	9.6 :+)
1.6	10	7.28	5.5	9.9
2.0	2.8	6.73	4.9	11.0
2.5	0.90	6.23	4.5	12.4
3.0	0.40	5.88	4.2	14.5
4.0	0.12	5.36		
6.0	$31 \cdot 10^3$	4.77	3.2	
8.0	13	4.40		
10.0	9.8		2.6	
15	2.5	3.68		
20	1.3		2.5	
30	0.46	2.94	2.24	

+) The first value in the last column may be affected by local turbulence.

should be approximately representative for average quiet conditions, the bulk velocities at 3 $R_\odot$ and 6 $R_\odot$ are $10^{6.5}$ and $10^{7.0}$ cm/sec respectively; for other distances it is easily derived from the condition of constant mass flow, if radial flow is assumed. We note already that out to at least $\approx$ 6 $R_\odot$ the flow should be subsonic. The last value (in the fourth column) is arrived at by assuming a further decrease of the density by a factor 80 between $r = 30\ R_\odot$ and 1 a.u. (for 460 : 80 = 5.7 cm^{-3} electrons, or 5.3 cm^{-3} ions; see the next section, I-C). In order to give an idea of the variability of the flux, the velocity and the density derived from them, we note the following: From data pertaining to the period from March 13 to June 1, 1968 (kindly made available by Dr. Dryer, ESSA) it appears that the average flux during the three solar rotations in question (1842-44) was only $\approx 0.8 \cdot 10^8\ cm^{-2}\ sec^{-1}$, the average velocity $\approx$ 500 km/sec, and the derived average density less than 2 cm^{-3}.

C. Macroscopic Properties of the Interplanetary Plasma Under Quiet and Disturbed Conditions. Methods of observations: Most direct measurements refer to the vicinity of the earth or, like the ones made e.g. during the flights to Venus and to Mars, to the vicinity of the ecliptic plane. Average values of the mass velocities--direction and magnitude--have also been obtained from the measurements of the plasma tails of comets, without any restriction to the ecliptic plane, out to $\approx$ 2 a.u. Finally radioastronomical observations--or the gradual occultation of point sources near the sun and the scintillation of quasi-stellar objects--have given further valuable evidence on the density fluctuations and the velocity field. Of the reviews listed in the first section, those of Axford[6], of Newkirk[3], and of R. Lüst[7] contain up-to-date accounts of and references to the observational material.

Velocity (direction, magnitude): The direction of the velocity is approximately radial, with fluctuations of usually a few, occasionally up to 10 or 20, degrees. Both from cometary observations[16] and direct measurements[17] there are indications of an azimuthal component, in the sense of co-rotation with the sun, of $\lesssim$ 10 km/sec. The amount of the velocity varies during quiet periods between $\approx$ 300 (lower limit $\approx$ 200 km/sec) and $\approx$ 400 km/sec up to 500 km/sec, with a strong tendency of recurrence; during disturbed periods, velocities up to 800 km/sec are found, during storms up to at least 1000 km/sec.

Chemical composition, state of ionization, density and flux: The average chemical composition appears to correspond to that of the sun's atmosphere[18], but with a percentage of α particles of only 4-5%; this latter varies considerably, particularly during interplanetary storms. The state of ionization corresponds to that of the inner corona, though the percentage of singly ionized helium is much higher (10^{-3} of He^{++} instead of 10^{-6}; a similar discrepancy has been found for O^{5+} relative to O^{6+} and O^{7+})[15]. For the average density reduced to 1 a.u., values between 5 and $\approx$ 7 (Pioneer VI) have been given in the literature. The fluctuations are rather large; during disturbed periods, values of up to 10^2 have been found, and, in storms, up to $2 \cdot 10^2$ (so far). The flux varies somewhat less, between 1 or $2 \cdot 10^8$ cm^{-2} sec^{-1} (lower limit $\approx 1/3 \cdot 10^8$) up to a few times 10^9 (so far).

Temperatures (ions and electrons): The temperatures are given here only as directional average of the microscopic kinetic energy; the microscopic velocity distribution will be discussed later with other evidence pertaining to microscopic phenomena. The ion temperature in this sense varies between several 10^4 degrees for very quiet conditions and 10^6 °K under disturbed conditions averaging $1.5 \cdot 10^5$ °K. The electron temperature is generally higher than the proton temperature and apparently never smaller than 1 to $2 \cdot 10^5$ °K [23] (cf. also section V).

Recurrent and flare-caused storms: The velocity and the density of the interplanetary plasma exhibit the tendency to recur with the period of the sun's rotation known from geomagnetic disturbances. In addition there are interplanetary storms caused by individual events on the sun, normally chromospheric eruptions of "flares", which emit a considerable quantity of plasma with a velocity of $\approx$ 1000 km/sec; the total energy released may be as high as 10^{32} erg or 10^{41} keV (with, say 10 keV per particle). A quantitative picture of these storms is only beginning to emerge[19].

Magnetic fields: The large scale structure of the interplanetary magnetic field--see the recent review article by Dessler[20]--is characterized by the tendency of the lines of force to follow the Archimedian spiral defined by the sun's rotation and the radial outward bulk motion; the velocity given by the product of the sun's angular velocity of rotation and the distance of 1 a. u. is approximately equal to a typical plasma bulk velocity, and the so-called garden hose angle is therefore typically in the range of 30° - 60°. Whether an average small component of the field perpendicular to the sun's equatorial plane exists, is not known; if it did, it would be hard to understand, since it would involve a steady transport of magnetic flux[+)]. Since there can be

---

+) In a discussion after the lecture, the question was raised whether such a transport for limited periods of time and with the sign changing e. g. each time could not be present. In this connection, attention was drawn to L. Biermann and R. Lüst's work[24] and to later work by Winge, Coleman et al[25]. (cf. section IV-a below).

no net magnetic flux into or out of the sun, there must be a balance between interplanetary lines of force directed towards the sun and away from it. Ness and Wilcox[21)] found a rather long-lived regular pattern in the interplanetary magnetic field characterized by only four large magnetic "sectors" (two of each sign), which co-rotated with the sun and matched in their polarity and extent the large scale features ("magnetic regions" with relatively weak fields) visible on the magnetograms obtained by optical means. Comparing the magnitude of the solar fields with those measured in interplanetary space (5-10 γ; geometrical factor $4.5 \cdot 10^4$ at 1 a.u.) it is seen that a considerable part of the flux of such regions is indeed carried by the solar wind into interplanetary space. On the other hand, lines of force belonging to regions of strong solar fields are ordinarily not carried into interplanetary space. Magnetic fluctuations may be interpreted as (e.g. magnetohydrodynamic) waves propagating in the plasma, of which several types would be expected to exist under the given circumstances (see sections III-C and III-D).

References of Section I[+)]

1) D. Billings, A Guide to the Solar Corona, Academic Press, New York - London, (1966).

2) H. Zirin, The Solar Atmosphere, Blaisdell, Waltham, Mass. - Toronto (1966).

3) G. Newkirk, Ann. Rev. Astron. and Astrophys. 5, 213 (1967).

4) R. N. Thomas, Aerodynamic Phenomena in Stellar Atmospheres, IAU-Symposium No. 28, Nice (1965).

5) E. Parker, Interplanetary Dynamical Processes, Interscience Publishers, New York (1963).

6) W. I. Axford, Space Sci. Rev. 8, 331 (1968).

7) R. Lüst, Solar-Terrestrial Physics, Chapt. I, Academic Press, London (1967).

8) E. Parker, Research Frontiers in Fluid Dynamics, Chapt. 20, p. 677, Interscience Publ., New York - London (1965).

9) L. Spitzer, Physics of Fully Ionized Gases, Interscience Publ., New York - London (1962) 2nd rev. edition.

10) L. Goldberg, Ann. Rev. Astron. and Astrophys. 5, 279 (1967).

11) G. Athay, Ap. J. 145, 784 (1966).

12) L. Biermann, Z. Astrophys. 25, 161 (1948).
M. Schwarzschild, Ap. J. 107, 1 (1948); see also[1)].

13) J. C.James, Ap. J. 146, 356 (1966); see also[3)].

14) J. C. Brandt, Ap. J. 149, 447 (1967).

15) A. J. Hundhausen, H. E. Gilbert, S. J. Bame, Ap. J. 152, L3 (1968) and Los Alamos Preprint LA-DC-9584; cf. also B. Kozlovsky, CALTECH Preprint, May 1968.

+) In the following sections these references are referred to as I-1), I-2), I-3). . .

16) J. C. Brandt, M. Belton, M. Stevens, Astron. J. 71, 157 (1966)
J. C. Brandt, Ap. J. 147, 201 (1967)

17) I. Strong, J. Ashbridge, S. Bame, A. Hundhausen, Trans. Am. Geophys. Union 48, 182 (1967), and Internat. Symp. on Zodiacal Light and Interplant. Medium, Hawaii, January 1967.

18) L. Goldberg, E. Müller, L. Aller, Ap. J. Suppl. 5, 45 (1960)
Lambert (unpublished)

19) J. Wilcox, Lecture at the COSPAR Symp., Tokyo, May 1968 (Techn. Report Univ. Calif. Berkeley)

20) A. Dessler, Rev. Geophys. 5, 1 (1967)

21) N. F. Ness, J. Wilcox, Ap. J. 143, 23 (1966); Solar Physics 2, 351 (1967)

22) P. Coleman, Plan. Space Sci. 15, 953 (1967); J. G. R. Sp. Ph. 73, 61 (1968)

23) M. Montgomery, S. Bame, A. Hundhausen, Trans. Am. Geophys. Union 49, 262 (1968)

24) L. Biermann, R. Lüst, Astrophysica Norvegica IX, 61 (1964)

25) C. R. Winge, P. J. Coleman, L. Davis, Jr., E. J. Smith, D. E. Jones, Publication No. 651, Institute of Geophysics and Planetary Physics, UCLA, Jan 12, 1968.

II. HYDRODYNAMICS OF THE SOLAR CORONA AND THE SOLAR WIND

A. Hydrodynamics of the Solar Corona and the Solar Wind without Conduction, Viscosity, Magnetic Fields and Rotation. In this section we follow largely Parker's discussion I-5). The basic equations are those of the balance of mass and of momentum, the energy equation being replaced by some assumption on the relation between the temperature (T), the pressure (p), and the density (ρ), such as

$$T = \text{const}; \quad p \sim \rho \quad , \tag{1}$$

or the polytropic relation (the index o denotes a suitable reference level)

$$(p/p_o) = (\rho/\rho_o)^{\alpha} \quad , \tag{1^+}$$

or by an assumption on the dependence of T or ρ on radial distance based on observations (e.g. table I). With the perfect gas law (isotropic pressure)

$$p = (kT/\bar{m}) \cdot \rho \quad ,$$

we have from the equation for the flow of momentum under stationary conditions

$$\rho v \frac{dv}{dr} + \frac{dp}{dr} + G \frac{M_{\odot}}{r^2} \cdot \rho = 0 \tag{2}$$

the familiar Bernoulli equation (r is the radial distance from the sun's center, v is the mass velocity).

$$\frac{1}{2} v^2 + (kT/\bar{m}) \ln\rho - G \frac{M_{\odot}}{r} = \text{const.} \tag{3}$$

respectively

$$\frac{1}{2} v^2 + \frac{\alpha}{\alpha - 1} \frac{p_o}{\rho_o} \left(\frac{\rho(r)}{\rho_o}\right)^{\alpha-1} - G \frac{M_\odot}{r} = \text{const.} \qquad (3^+)$$

where $\alpha - 1$ is actually rather small-- 0.1 ... 0.2 --such that the second term varies only by roughly a factor 10 between coronal levels and the earth environment+).

The continuity of mass flow requires for $\underline{v} \| \underline{r}$

$$\rho v r^2 = \text{const.} \ (\approx 10^{11.0} \ \frac{g}{\text{sec,ster}}) \ . \qquad (4)$$

In the case of non-radial flow, r^2 has to be replaced by the cross section A of the tube of flow, which in general varies along a stream line. This formulation can also be used, as proposed by Clauser[1)], to elucidate the analogies with the flow through a Laval nozzle into a vacuum, the throat of the nozzle playing essentially the same role as the gravitational field in the solar case (for further details see Parker's monograph[I-5)], p. 63; or Dessler[I-20)]).

The various effects neglected in this treatment, especially the high value of the thermal conductivity (see sections II-B and II-C) result in a relatively slow variation of T with r between the corona and interplanetary space which, according to the observations (see sections I-C and I-B) amounts to only one or one and a half powers of ten in T against eight powers of ten for ρ. It is for this reason that we first fix attention to the isothermal case. A detailed discussion of the polytropic case, which does not differ qualitatively in any important way from that for T = const., has been given by Parker.

+) The equation expressing the transition from (3^+) to (3) is not helpful here, since $(\alpha - 1) \ln (\rho_o/\rho)$ is not $\ll 1$.

As long as $\nabla(1/2\ v^2) << \nabla(GM_{\odot}/r)$ (cf. section I-B), one obtains from equation (2) the familiar expression for the scale height.

$$\frac{1}{\nabla \ln p} = \frac{1}{\nabla \ln \rho} = \frac{kT/\bar{m}}{GM_{\odot}/r^2} = \frac{kT}{\bar{m}g} \quad . \tag{5}$$

Writing $v(dv/dr) = b$, we obtain from (2) the generally valid expression

$$\frac{1}{\nabla \ln p} = \frac{1}{\nabla \ln \rho} = \frac{kT/\bar{m}}{g + b} \quad ,$$

with the mean atomic mass $\bar{m} \gtrsim m_H/2 \approx 10^{-24.0}$ g (for $\approx 5\%$ He per volume) and with the Newtonian speed of sound $\sqrt{kT/\bar{m}}$ or $\approx \sqrt{2}$ times that of the ions. The scale height is the same for the ions and the electrons. This reflects the influence of the electric surface field, which, with an exceedingly minute excess of ions over electrons, diminishes the weight of the ions by $\approx 1/2$ and gives effectively this weight to the electrons; the result is that the effective escape velocity for both is only $\sqrt{1/2}$ of that of non-ionized atoms (cf. section III-b), or 437 km/sec instead of 618 for $r \approx R_{\odot}$.

The value of $(kT/\bar{m})$ for average conditions in the equatorial belt is $\lesssim 10^{14.4}$ $(cm/sec)^2$; in active regions it may be more than twice as high, in inactive polar regions as low as $10^{14.2}$ $(cm/sec)^2$. The gravitational acceleration $GM_{\odot}/r^2$ is $2.74 \cdot 10^4$ cm/sec^2 at the surface, where $r = R_{\odot} = 7.0 \cdot 10^{10}$ cm.

Returning to the Bernoulli equation for the isothermal case, we now introduce the dimensionless variables (using Parker's notation)

$$U = \frac{v}{\sqrt{2kT/\bar{m}}}$$

$$\pi = r \cdot \frac{kT/\bar{m}}{GM_\odot} \quad \left(\geq R_\odot \frac{kT/\bar{m}}{GM} \approx 10^{-0.9} \right) \tag{6}$$

which yield (using (4))

$$U^2 - \ln U = 2 \ln\pi + \frac{1}{\pi} + L \tag{7}$$

where L is the constant of integration. For large distance from the sun (large π) we have asymptotically either

$$U^2 \approx \ln\pi^2 \qquad (U^2 \text{ large}) \tag{8}$$

or

$$U \sim \frac{1}{\pi^2} \sim \frac{1}{r^2} \qquad (U^2 \text{ small}) \tag{9}$$

and for π small

$$U^2 \approx \frac{1}{\pi} \sim \frac{1}{r} \rightarrow \frac{1}{R_\odot} \qquad (U^2 \text{ large}) \tag{10}$$

or

$$\ln\left(\frac{1}{U}\right) \sim \frac{1}{\pi} \qquad (U^2 \text{ small}) \tag{11}$$

By differentiation we obtain the relation

$$\left(2U - \frac{1}{U}\right) \frac{dU}{d\pi} = \frac{1}{\pi} \cdot \left(2 - \frac{1}{\pi}\right) \tag{12}$$

which shows that $dU/d\pi$ becomes indeterminate at the "critical" point

$$U = \frac{1}{\sqrt{2}} \quad , \quad \pi = \frac{1}{2} \quad , \quad L = 0.232$$

There we have

$$v = \sqrt{\frac{kT}{\bar{m}}} \qquad \text{(Mach number 1, referred to Newton's velocity of sound)} \tag{13}$$

and

$$\frac{3}{2}\, kT + \frac{1}{2}\, \bar{m}v^2 = \bar{m}G\, \frac{M_\odot}{r} \quad ,$$

the individual terms being in the ratio 3 : 1 : 4.

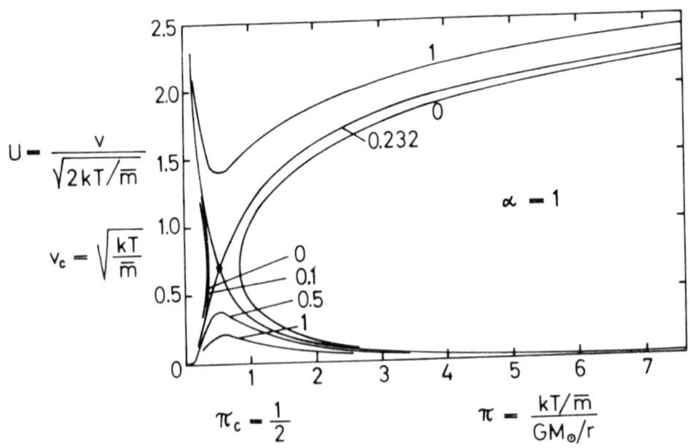

Note that at the critical point $(1/2)\bar{m}v^2$ is only one quarter of the potential energy. Of the two solutions for large distance, the second one leads with the equation of continuity to constant and finite values of density and pressure (with

v^2 tending towards zero); this evidently is the wrong boundary condition for interplanetary space in view of the very small value at the outer boundary (10^{-12} dyn/cm^2). This is true also if $(kT/\bar{m}) > GM_\odot/2R_\odot$ though in this case no stationary solution may be possible (if the temperature is artificially maintained at that level; in the limiting case of effectively zero gravity, as on the surface of cometary nuclei, only hypersonic solutions are possible). The first solution (cf. equation (8)), on the other hand, has the correct behavior.

Turning to the other extreme, π small, we see similarly that only the second one fits the general conditions given in the solar corona, though in this case the applicability of the analysis to the physical situation is obviously limited.

In the case of the sun, the critical point $\pi = 1/2$ is for the isothermal corona discussed here at

$$r = \frac{1}{2} \frac{GM_\odot}{kT/\bar{m}} \lesssim 4R_\odot \approx \frac{2/3 \cdot 10^{26}}{10^{14.4}} \approx 10^{11.4} \text{ cm}$$

roughly (in view of the approximations used). At that point, $\lesssim$ 75% of the gravitational potential of the sun has been overcome already, but a similar amount of energy must flow beyond r_c for accelerating the plasma to its final velocity.

We note in passing that, though we used the formulae given here to describe the outflow of the coronal gas into interplanetary space, they could be used just as well to describe the inflow of interplanetary gas into the solar atmosphere. This case had been treated much earlier, but the relationship between the two cases was noted only after Parker's work.

The appearance of the singularity at the critical point is of course due to the neglect of viscosity. Axford and Newman[2)] have shown in detail how the inclusion of viscosity and conduction removes this singularity and leads to the appearance of weak non-Rankine-Hugoniot shock waves in the solutions.

Note: The quantities U, π may also be defined locally (e.g. ≈ 6 and ≈ 3 near 1 a.u.). If a more accurate approximation is sought by joining several isothermal zones, the conditions of fit are (cf. (4) and (2), with $(GM/r^2)\bar{\rho}\Delta r \ll \Delta p$ and $\rho v \Delta v$ across the boundary).

$$(n\bar{m}v)_i = (n\bar{m}v)_{i-1} = (n\bar{m}v)$$

and with

$$(n\bar{m}v)v_i + kT_i n_i = (n\bar{m}v)v_i + (kT)_i \frac{(n\bar{m}v)}{\bar{m}v_i}$$

$$= (n\bar{m}v)v_{i-1} + kT_{i-2}n_{i-1}$$

$$v_i + \frac{(kT/\bar{m})_i}{v_i} = v_{i-1} + \frac{(kT/\bar{m})_{i-1}}{v_{i-1}} \quad .$$

Putting

$$\frac{v}{\sqrt{kT/\bar{m}}} = V = \sqrt{2}\, U,$$

this is

$$\left(\sqrt{\frac{kT}{\bar{m}}}\right)_i \left(V_i + \frac{1}{V_i}\right) = \left(\sqrt{\frac{kT}{\bar{m}}}\right)_{i-1} \left(V_{i-1} + \frac{1}{V_{i-1}}\right)$$

V^{-1},	0.1	0.16	0.25	0.4	0.63	1
V,	10	6.3	4	2.5	1.59	1
$V + V^{-1}$	10.1	6.5	4.25	2.9	2.22	2

A factor $\frac{1}{\sqrt{2}}$ in the $\sqrt{\frac{kT}{\bar{m}}}$ term gives rise to a factor $\sqrt{2}$ in the function $(V + \frac{1}{V})$, that is, to a factor somewhat larger than $\sqrt{2}$ in V itself, which is effectively the Mach number, whereas the scale of π changes (contracts) in the ratio of T itself (or a factor 1/2 in the example).

If, as will be discussed later, T is $\sim r^{-1/2}$, π would become $\sim r^{1/2}$. If furthermore, as seems to be true for $r > 30\ R_\odot$ or so, v increases only slowly with r, say $\sim r^{1/4}$ on the average, U would also become $\sim r^{1/2}$. At large distances U should therefore be approximately $\sim \pi$.

B. <u>Hydrodynamics of the Solar Corona, Including Thermal Conduction and Viscosity</u>. If only thermal conduction is included, that asymptotic solution at large distance, which is characterized by small mass velocity, now has values of the density and the temperature, which may go to zero and hence cannot be excluded a priori on the ground of a physically impossible boundary condition. At the time just before the measurements from space vehicles removed the last doubt about the actually existing physical situation, there was an argument in the literature about which of the asymptotic solutions would give the more correct description of the conditions in interplanetary space (cf. e.g. reference I-20). We shall not discuss this case, which has been treated by

Chamberlain, by Scarf and Noble, by Parker+), by de Jager, and by Whang and Chang, since it has become apparent subsequently that conduction and viscosity should be included simultaneously (their contributions to the energy equation in the critical regions being of similar order as will become apparent below). The case with normal isotropic viscosity and with isotropic heat conduction has been worked out by Whang, Liu and Chang[3]; as will be seen below, the more realistic assumption of an anisotropic viscosity and conductivity in the presence of the interplanetary magnetic field $\underline{B}$ does not lead to different results as long as the angle between $\underline{r}$ and $\underline{B}$ is not too large. We follow essentially the notation of Whang, Liu and Chang, assuming, as before, spherical symmetry, and discuss first their procedure and results.

The three equations of conservation of mass, momentum and energy are

$$\rho V r^2 = m = \text{const.} \tag{1}$$

$$V \frac{dV}{dr} = -\frac{1}{\rho}\frac{dp}{dr} - G\frac{M_\odot}{r^2} + \frac{1}{\rho r^3}\frac{d(r^3\tau)}{dr} \tag{2}$$

$$m\left(\frac{3}{2}a^2 + \frac{1}{2}V^2 - G\frac{M_\odot}{r}\right) + r^2 q - r^2\tau V = mh_o \tag{3}$$

with $a^2 = 5\,kT/3\bar{m}$. We note already $m \approx 10^{11.0}$ g/sec/ster

---

+) E. Parker (Ap. J. 141, 1464 (1965)) has used these equations (viscosity = 0, $T_i = T_e$) to estimate two critical coronal densities N_A and N_B (at given temperatures T_o) below which there is essentially conductive transport of energy (N_A) respectively above which (N_B) the expansion becomes subsonic, the conductive heat flow maintaining the steady expansion being given essentially by T_o alone (cf. also Brandt et al., ICARUS 4, 19 (1965).

(or $10^{35.0}$ ions + electrons) under generally quiet conditions and $h_o \approx 0.4\ GM_\odot/R_\odot \approx 0.8 \cdot 10^{15}$ erg/g for a mass velocity at large solar distance of ≈ 400 km/sec (the terms containing q and τ being there small against $mV^2/2$, cf. V-A).

For the viscous stress and the thermal conduction we have

$$\tau = \frac{4}{3}\mu r \frac{d}{dr}\left(\frac{V}{r}\right) = \frac{4}{3}\mu\left(\frac{dV}{dr} - \frac{V}{r}\right) \frac{\text{dyn}}{\text{cm}^2} \tag{4}$$

and

$$q = - K\frac{dT}{dr} = - \frac{GKa}{5(k/\bar{m})} \frac{da}{dr} \tag{5}$$

with $\mu = 10^{-16.0} \cdot T^{5/2}$ and $K = 10^{-6.0} \cdot T^{5/2}$ for the solar corona, of $10^{-16.1} \cdot T^{5/2}$ and $10^{-6.1} \cdot T^{5/2}$ respectively in interplanetary space ($n_e \leq 10^5$ cm^{-3}, cf. [I 9]), where we have assumed $\lesssim 5\%$ He by volume. Both values refer to conditions in which, as was assumed by Whang et al., the influence of the magnetic field can be neglected; we shall come back to this point later. As a matter of fact, the authors assume 10% He and treat μ/K as an adjustable parameter.

The equations (1) - (5) are now reduced to a system of three differential equations for the unknown V, a (instead of T), and τ.

$$\left(V^2 - \frac{3}{5}a^2\right)\frac{dV}{dr} + \frac{6}{5}aV\frac{da}{dr} - \frac{V^2r^2}{m}\frac{d\tau}{dr} = \frac{Ga^2V}{5r} -$$

$$- G\frac{M_\odot}{r^2}V + \frac{3V^2\tau r}{m} \tag{6}$$

$$\frac{dV}{dr} = \frac{V}{r} + \frac{3\tau}{4\mu} \tag{7}$$

$$\frac{da}{dr} = \frac{5}{6} \frac{m(k/\bar{m})}{Kar^2} \left(\frac{3}{2} a^2 + \frac{1}{2} V^2 - G \frac{M_\odot}{r} - \frac{r^2 \tau V}{m} - h_o \right) \tag{8}$$

After introducing non-dimensional variables and parameters, of which we note here (the $*$ denotes conditions at a reference sphere)

$$Z = \frac{r}{r*} \quad , \quad u = \frac{V}{V*} \quad , \quad A = \frac{a}{V*} \quad , \quad W = \frac{\tau}{\rho V*^2}$$

and

$$M = \frac{u}{A} = \frac{V}{a} \quad , \quad R_e^* = \frac{\rho * V * r *}{\mu *} \quad , \quad P_r = \frac{5}{2} \frac{k}{\bar{m}} \frac{\mu}{K} \quad ,$$

$$\gamma = \frac{GM_\odot / r*}{V*^2} \quad .$$

With these substitutions, equations (6) - (8) are written as follows

$$a_{i1} \frac{du}{dZ} + a_{i2} \frac{dA}{dZ} + a_{i3} \frac{dW}{dZ} = a_{i4} \tag{9}$$

The values of the a_{ij} and of the resultant determinants will not be given here; we note only that the denominator of the system, the determinant

$$Q_{123} = \frac{3M*^5}{P_r R_e^*} Z^4 u^2 A^6$$

does not vanish for any permitted value of Z (Pr is of the order of 1/10).

Whang, Liu and Chang choose their final solution on the basis of certain plausible limiting conditions, which are then used to deduce the only physically meaningful solution. They consider that at infinity u should remain finite, but both the pressure and the temperature--even more the coefficients of viscosity and conduction, which are $\sim T^{5/2}$ --should tend to zero. Writing $A \sim Z^{-m}$ and $W \sim Z^{-n}$, they show then that m must be $> 1/5$ and $n > 2$, that at large distance all the energy is transported by convection of kinetic energy, and also that the work done against the viscous stresses tends towards zero.

They choose 1/4, 1/3, and 1/2 as trial values for m and find that only $m = 1/3$ (with $n = 8/3$) leads to a physically meaningful solution, which is worked out for $Z > 1$ in the form of power series of Z^{-m}. They find it useful to make the radius of the reference sphere and M^* quite large; they choose $M^* = 15$ for rapid convergence of the series, and $r^* = 1.8 \cdot 10^4\ R_\odot$ (one of the two choices being arbitrary). The computations are carried out for four values of the Prandtl number Pr between 1/12 and 1/60. Though kinetic theory suggests that the first or the second value (1/20) should be best, it is the last one which ultimately gives the best fit with the coronal observations (cf. e.g. table 1). μ_∞ and R_e^* are found from the power series as functions of Pr and γ. The general run of the solutions of the system (9), with γ as a parameter in the u,A plane shows that there are two families, one with u becoming negative, the other with M increasing toward the sun; only the solution dividing these two families is physically acceptable. This fixes γ as a function of the Prandtl number ($2.74 \cdot 10^{-4}$ ... $2.96 \cdot 10^{-4}$ in the range of Pr considered). Since the transport terms become small far from the sun, it is not surprising to rediscover there a similar behavior as in the former discussion (note that the family of solutions with $u \to 0$ has been excluded beforehand).

The value of γ found from the solution with the correct behavior at small Z yields V^{*2} (that is h_o in equation (3)), which is 189 km/sec for $Pr = 1/60$ (for $Pr = 1/12$ it is 4% larger), or 165 km/sec at 1 a.u., too low by a factor ≈ 2. This is largely due to the fact that the viscous term in the momentum equation becomes negative (decelerating) beyond $r \approx 7\ R_{\odot}$. The electron density is $10^{-3.0}$ at $Z = 1$, or $\approx 10^{0.9}\ cm^{-3}$ near 1 a.u., such that the flux is $10^{8.1}\ cm^{-2}$ sec, approximately as observed. The temperature drops from $1.6 \cdot 10^6$ in the corona to $0.9 \cdot 10^5$ at 1 a.u. The coronal densities are in reasonable agreement with the observed ones, but only by choosing an, at first sight, improbable value for the Prandtl number.

The discrepancy by a factor of 3 or 4 in the parameter Pr could perhaps, since μ is $\sim T_{ion}^{5/2}$, K however $\sim T_e^{5/2}$, be accounted for by an average value for $T_e/T_{ion} \approx 1.6$, which is not implausible (cf. I-3).

There are three major comments which might be made concerning this work: (1) The interplanetary magnetic field will affect the transport both of momentum and of heat at least at large distances from the sun. (2) In view of the temperature differences between ions and electrons, which are now known to exist, the macroscopic equations should be written down individually for the ions and the electrons. This has been done by Hartle and Sturrock (see next section II-B). (3) If the average temperatures actually measured in interplanetary space are taken, it is found, as will be discussed in section III-A, that the macroscopic equations can legitimately be used, for the ions at least, only near the sun (though secondary, e.g. microscopic effects may keep the deviations from the macroscopic picture within certain limits). It is therefore questionable whether for $r > 7 R_\odot$ the diminution of the acceleration found by Whang et al. (see above) will really occur. This subject will be pursued in section III-D.

As to the first point, the anisotropy of the viscosity due to the magnetic field can be shown not to reduce the viscous acceleration or deceleration as was re-emphasized by Busemann[7], cf. Meyer and Schmidt[8),4) +)]. The anisotropy of the thermal conductivity, which is greatly reduced $\perp \underline{B}$ in interplanetary space, means essentially a reduction by the cosine of the angle between $\underline{B}$ and $\underline{r}$ (cf. section IV), which is not too important for $r < 1$ a.u. The radial momentum equation with the magnetic field term will be given in section IV.

We show finally that a consistent picture of the integral properties of the solar corona and those found empirically near 1 a.u. (cf. Brandt et al, loc.cit. footnote p. II-24) may be arrived at by using simply the equations (1) - (3) with the constants given already. $VdV/dr \ll g = GM/r^2$ and $\tau \ll p$ in the corona itself permit us to derive $(kT/\bar{m}) = g/(\nabla \ln N_e + \nabla \ln T)$, which is approximately proportional to the inverse of the last column of Table I. $\nabla \ln T$ being only a correction (of the order of 10% below)

+) The viscous acceleration and particularly the viscous deceleration is, however, considerably reduced by the anisotropy of the microscopic velocities[8)], which will be discussed in section IV.

$3\ R_\odot$, it is easy to derive $kT/\bar{m}$ from the observed densities and thus also $\nabla \ln T$; the "temperature" thus determined should of course include turbulence for $r < 1.6\ R_\odot$. From Newkirk's Table I we find $d(\ln T)/d(\ln r) \approx -0.5$ between $r = 1.6\ R_\odot$ and $3\ R_\odot$ ($\Delta \log T \approx 0.10$ over $\Delta \log r \approx 0.19$). It can then be verified that (with $\bar{m} = 10^{-24.0}$) $T = 10^{6.1}$ at $1.8\ R_\odot$ is consistent with the assumed value of h_o and the heat flux required to provide the energy for overcoming gravitation and accelerating the solar wind still further:

$$m(GM_\odot/R_\odot)\left[\frac{h_o}{GM_\odot/R_\odot} + \frac{R_\odot}{r} - \frac{5}{2}\frac{kT}{\bar{m}}\Big/\frac{GM_\odot}{R_\odot}\right]$$

$$= - rTK\frac{d(\ln T)}{d(\ln r)} \approx 1.4 \cdot 10^{26}\ \frac{\text{erg}}{\text{sec,ster}}$$

Here mh_o may be said to be known from the measurements near 1 a.u., $T/\bar{m}$ and $d(\ln T)/d(\ln r)$, however, from the coronal observations of the densities, and independently of the spectra. Inside of $r = 1.6\ R_\odot$ the temperature appears to rise more slowly in agreement with the radar observations, indicating strong heating by oscillations or shock waves out to about that distance; also radiative cooling (neglected in the work discussed in this section) should be important there.

C. <u>Two-Temperatures Model of the Solar Wind</u>. The exchange of (thermal) energy between electrons and ions is slower than that between electrons of different energy (in the range of thermal energies) in approximately the ratio of their masses. As a consequence, it is also slower than that between ions of different energy. The first deviation from equilibrium conditions of an element of mass of the solar corona participating in the outward motion, to become part of the solar wind, must therefore be the appearance of different temperatures of the ions and the electrons, long before the rareness of collisions leads to anisotropies of the pressure of one or of the other. The first stages of the development of this temperature difference are therefore adequately described by the following set of equations (Braginskii 1965[5)]) which were used by Hartle and Sturrock (1968)[6)] in their discussion of the solar wind problem; in this work the influence of the viscosity has been neglected (notation essentially as in section II-B, the indices referring to electrons and protons respectively since other ions were neglected; n, however, $= n_e = n_i$ in this section).

The energy equations for the ions and the electrons are used in the form (cf. 5))

$$(nVr^2)(kT_e)\left(\frac{3}{2}\frac{d(\ln T_e)}{dr} - \frac{d(\ln n)}{dr}\right) - \frac{d}{dr}\left(r^2 K_e \frac{dT_e}{dr}\right) = -\frac{3}{2} n\nu_E r^2 (kT_e - kT_i) \frac{\text{erg}}{\text{sec,cm}}$$

$$(nVr^2)(kT_i)\left(\frac{3}{2}\frac{d(\ln T_i)}{dr} - \frac{d(\ln n)}{dr}\right) - \frac{d}{dr}\left(r^2 K_i \frac{dT_i}{dr}\right) = +\frac{3}{2} n\nu_E r^2 (kT_e - kT_i) \frac{\text{erg}}{\text{sec,cm}}$$

with

$$K_e = 3.16\ k\,\frac{nkT_e}{m_e V_e} \qquad K_p = 3.9\ k\,\frac{nkT_p}{m_p V_p}$$

(only protons being taken into regard by Hartle and Sturrock)

$$\nu_E = 2 \frac{m_e}{m_p} \nu_e$$

$$\nu_e = \frac{4}{3} \sqrt{2\pi} \left(\frac{kT_e}{m_e}\right)^{1/2} \frac{ne^4}{(kT_e)^2} \ln \Lambda \quad ;$$

$$\nu_p = \frac{4}{3} \sqrt{\pi} \left(\frac{kT_p}{m_p}\right)^{1/2} \frac{ne^4}{(kT_p)^2} \ln \Lambda$$

$$\Lambda = \frac{3}{2\sqrt{\pi}} \left(\frac{kT}{e^2}\right)^{3/2} n^{-1/2}$$

Finally, we note that the momentum equation now contains the gradient of $nk(T_e + T_p)$ instead of that of 2nkT; charge neutrality ensures that in the macroscopic description it is unnecessary to write two separate momentum equations for the ions and the electrons. We note, furthermore, that $T_e >> T_i$ would mean $p_e >> p_i$ and $m_p(V(dV/dr) + g) = eE$ (cf. section III-B below) $= -(1/n)dp_e/dr)$. If the relatively slow variation of T is neglected, this equation becomes

$$eE = -\frac{1}{n} \nabla p_e = -(kT_e)_{av.} \cdot \nabla \ln n$$

Without the right-hand-side term of the heat equations (p. 31), which expresses the exchange of thermal energy between the electrons and the protons, the large difference of the thermal conductivities would result in a much slower drop of T_e with increasing r than of T_i. The results of Hartle and Sturrock show that, even with the exchange term, considerable temperature differences may develop at larger solar distance (10 $R_\odot$ and more).

The neglect of the viscous terms in the momentum and energy equations has the consequence that, as in the isothermal case, a "critical" point appears in which dV/dr is indeterminate and through which the solution with the desired properties at large distances must pass. This condition is one of the six which determine the solution of the system (including the continuity and the momentum equation).

Four conditions refer to the temperatures T_p and T_e at the base (T_{po} and T_{eo}) and at large distance (= 0 in the work of Hartle and Sturrock). Finally the value of n at the base ($n(r_o) = n_o$) was used as the sixth condition.

Since the heat equations are certainly incomplete for $r \lessapprox 1.7\ R_\odot$, the conditions near $r = 2R_\odot = r_o$ are the ones which should perhaps be taken as reference values. One model preferred by the authors, owing to the good fit with Blackwell's data of the 1954 eclipse has $T_{eo} = 1.5 \cdot 10^6$, $T_{po} = 1.2 \cdot 10^6$, and $n_o = 1.5 \cdot 10^6$ at $r_o = 2R_\odot$; the integration had actually been started at $r \approx R_\odot$ with $T_p = T_e = 2 \cdot 10^6$ and $n = 3 \cdot 10^7$. That a substantial temperature difference $T_e - T_p$ may already be present at $r = 2R_\odot$ can of course be checked also from $V(r)$ and the values of the temperature conductivities derived from the K's.

Near 1 a.u., the computed values are $n \approx 15\ \text{cm}^{-3}$ and $V \approx 250$ km/sec, the first of which is larger than the measured average value, while the second is on the low side. T_e and T_p (1 a.u.) are found to be $3.4 \cdot 10^5$ and $4.4 \cdot 10^3$ respectively, the first of which is probably somewhat too high, while T_p is definitely too low.

The time scales ν_p^{-1} and $\nu_E^{-1} \approx 10^{+3.0}\ \nu_e^{-1}$ may be compared with the expansion time scale

$$\left(\frac{d(\ln n)}{dt}\right)^{-1} = \left(\frac{v}{r}\frac{d(\ln n)}{d(\ln r)}\right)^{-1} .$$

If (as done by the authors) the computed values of T_p and V (1 a.u.) are used instead of the observed ones, ν_p is found to be $\gg V_{exp.}$; while V_E is $< V_{exp.}$ from $\approx 5\ R_\odot$ outward. The differences between the computed results concerning $T_e(r)$, $T_p(r)$, and $n(r)$ on one hand and the observed values on the other are interpreted by the authors as indicating that additional heating of the protons goes in the corona at $r > 2\ R_\odot$ and actually out to at least $\approx 200\ R_\odot$ (1 a.u.).

The discussion of these time scales will be taken up again in the next section.

References of Section II

1) F. Clauser, Fourth Symp. on Gas Dynamics, IAU-Symp. 12, p. 26 (1961).

2) W. Axford and R. Newman, Ap. J. 147, 230 (1967).

3) Y. Whang, C. Liu, C. Chang, Ap. J. 145, 255 (1966).

4) F. Meyer, H. U. Schmidt, Mitt. Astron. Ges. 21, 97 (1966).

5) S. Braginskii, Rev. Plasma Phys. 1, 205 (1965).

6) R. Hartle, P. Sturrock, Ap. J. 151, 1155 (1968).

7) A. Busemann, personal communication, (1968).

8) F. Meyer, H. U. Schmidt, Mitt. Astron. Ges. 25, 198 (1968) (correction to 4).

III. KINETIC THEORY OF THE SOLAR CORONA AND SOLAR WIND

A. <u>On the Different Levels of Theoretical Description</u>. If collisions are frequent, that is to say if the inverse collision frequency is small compared to all time scales of interest in connection with the dependence on space and time of the macroscopic quantities, it is easy to derive from the Boltzmann and Maxwell equations a macroscopic model of a fully ionized gas (cf. III-B). For a chemically uniform gas these are the so-called two-fluid equations. As will be shown in the next section, they are approximately true even when the gyration frequencies of the charged particles are large compared to the collision frequencies. For densities so low that not only do the electrons and ions gyrate many times before colliding with another charged particle, but that also the condition specified above is not fulfilled, the "microscopic" method of the Boltzmann equation (including of course Maxwell's equations) or its limiting form for no collisions at all, the Vlasov equation, should properly be used; these do not seem to have been applied so far to the problem of the large scale properties of the interplanetary plasma (see, however, the addendum III-D).

Intermediate levels of description using the properties of certain invariants of motion (e.g. the so-called guiding centre approximation in which different "temperatures" parallel and perpendicular to the magnetic lines of force may be introduced) have been given by Chew, Goldberger and Low[1] and by Thompson[2] and have been applied e.g. to the problem of the occurrence of certain instabilities and of shock fronts, in the structure of which the gyration radii of the charged particles seem to appear[+].

[+] The Liouville equation, when applied to such problems, may be said to be equivalent to a certain way of using the Vlasov equation, as has been discussed first by Y. L. Klimontovich, JETP <u>6</u>, 753 (1958).

It should be observed, however, that the general agreement between the measured large scale structure of the interplanetary magnetic field and the predictions based on the simplest form of the macroscopic theory at least appears to indicate (though the reasons for this are by no means clear) that no very large errors are made by using, for normal conditions, the magnetohydrodynamic picture. Under disturbed conditions, especially during interplanetary storms caused by solar eruptions, the kinetic collision frequency becomes very much smaller while, at the same time, the magnetic fields may become rather irregular. As a result, the macroscopic equations may indeed lead to considerable errors under such circumstances.

In order to derive the limits in question under general conditions, as given by the kinetic theory of plasmas, we note (cf. Spitzer[I-9], chapter 5) that the self-collision time for protons is

$$t_c = \frac{11.4}{\ln \Lambda} \cdot \frac{T^{3/2}}{n} \quad \text{sec}$$

with Λ being given by the relation (λ_D is the Debye shielding distance).

$$\Lambda = \frac{\lambda D}{e^2/kT} = \frac{(kT)^{3/2}}{n^{1/2} e^3} \cdot \frac{3}{2\sqrt{\pi}}$$

since

$$\lambda_D = \left(\frac{kT}{4\pi n e^2}\right)^{1/2} = 6.9 \cdot \sqrt{\frac{T}{n}} \quad .$$

Λ is seen to be essentially the number of particles in a Debye volume $n\lambda_D^3$. Thus there are three scales of length in a plasma with no magnetic field, the "Landau length"

$$\frac{e^2}{kT} \approx 1 \text{ Å} \quad \text{for} \quad kT \approx 15 \text{ eV}$$

$$(T \approx 1.7 \cdot 10^5 \ {}^\circ K),$$

the average distance

$$n^{-1/3} = 0.6 \text{ cm} \quad \text{for} \quad n = 5 \text{ cm}^{-3},$$

and the "Debye length"

$$\lambda_D = 1.4 \cdot 10^3 \text{ cm} \quad \text{for} \quad T = 2 \cdot 10^5, \quad n = 5 \text{ cm}^{-3}.$$

In volumes $< \lambda_D^3$, the fluctuations of the numbers of the ions or of the electrons can each be shown to be essentially statistical, whereas, for volumes $>> \lambda_D^3$, the net electric charge remains always very small.

In the equation for t_e given above, the term $\ln \Lambda$ takes into account the effect of distant encounters and is given by the following table (loc.cit., p. 128), which pertains to an electron-proton plasma:

T	$n_c = 10^0$	10^3	10^6	10^9 cm^{-3}
10^4	23.2	19.7	16.3	12.8
10^5	26.7	23.2	19.7	16.3
10^6	29.7	26.3	22.8	19.3

In order to take the collisions with α particles (perhaps $\lesssim 5\%$ of the protons) into account approximately, we use the round figure of 10 instead of 11.4. For electrons we have a coefficient $\sim A^{1/2}$ (A = atomic weight), or ≈ 0.27 instead of 10. The effective cross section is $\sim T^{-2} \ln \Lambda$, as is easily verified. If the temperature variation with solar distance were smooth, a law $T \sim r^{-1/2}$ would fit the average data both in the corona and near 1 a.u. for the electrons. Though it is doubtful whether this is an adequate representation for the ions also, we shall make use of it in the absence of more reliable information.

We use the equation for t_c now for checking whether or not the condition stated in the beginning of this section is fulfilled in interplanetary space. There we are at 1 a.u., only under very quiet conditions (low ion temperature, somewhat higher density), just at the limit of the applicability of the macroscopic equation, the inverse collision frequency ($10^{5.4}$ sec for $T = 10^{4.5}$, $N = 8\ cm^{-3}$) of the protons being of the same order as the time in which the density or pressure change materially due to the convection with the solar wind (0.5 a.u. with 300 km/sec, also $2.5 \cdot 10^5$ sec). Normally, however, the temperature is higher ($10^{5.2}$, Axford[I-6]) and the density lower ($5\ cm^{-3}$ or even less), such that the inverse ion collision frequency becomes $10^{6.7}$ sec. Since the bulk velocity is also higher ($4\text{-}5 \cdot 10^7$ cm/sec), the time of a change of n by a factor e by convection is now only $10^{5.2}$ sec. Only for the electrons, for which the self-collision frequency is higher in the ratio $(m_p : m_e)^{1/2} \approx 43$ and the thermal speed ($\approx 10^{8.3}$ cm/sec) is larger than that of the bulk motion, the discrepancy is moderate. (The velocity together with t_c yields a mean free path of $10^{13.4}$ cm, which is still approximately halved by the collision with ions; the effective collision cross section is found to be $\approx 10^{-14.1}\ cm^2$. The distances actually covered between two collisions which should be compared with $(|\nabla \ln n|)^{-1}$ are further reduced somewhat by the spiralling around the magnetic lines of force.)

To the extent to which the bulk velocity is constant, the density must vary with solar distance $\sim r^{-2}$. The evidence on the correlation between the polarities of the magnetic fields on the sun and near the earth discussed in section I-C (cf. also [3)]) shows that the average radial solar wind velocity between the earth and the sun in December 1963 and the first two months of 1964 was 385 ± 45 km/sec, assuming there was no co-rotation of the corona (effective co-rotation out to 21 $R_\odot$ = 0.1 a. u. would diminish the velocity by 10%), whereas the measured bulk velocities for the same period of time from two different instruments were 360 and 378 km/sec[3] respectively. These values are quite close together; we conclude that over most of the distance between sun and earth the bulk velocity should not be too different from the value observed in our vicinity. As a consequence, also the density should approximately vary $\sim r^{-2}$ over the same region (Newkirk's value of the density at 30 $R_\odot$ with Axford's value of 5 cm^{-3} at 1 a. u. would give a discrepancy by a factor $\lessapprox 2$ in the ratio of the two values, which should, however, still be in the range of the combined uncertainties of the two figures). The influence of the variation of T and n on the mean free path λ are thus seen to be such that λ becomes approximately $\sim r$. The times between two collisions are then $\sim rT^{-1/2}$ or $\sim r^{5/4}$. On the other hand, the time scale of the change of n through convection, $(|\nabla \ln\rho| v)^{-1} \approx r/2v$, is seen to decrease towards the sun for the ions approximately as r^1 (the somewhat steeper increase of n inwards and the decrease of v acting in opposite directions). That is to say that the discrepancy of the two times t_c and $(v\nabla \ln\rho)^{-1}$ at $r = 30\ R_\odot$ is, for the ions, almost as large as at 214 $R_\odot$ (1 a. u.), as may be checked also directly. For the ions the conditions stated above are hence nowhere fulfilled under normal conditions (for $T \lessapprox 10^5$).

For the electrons the expression for t_c is the same, apart from the factor $10^{-1.6}$ stated above (of course the frequency of collisions with ions is approximately the same). We have now to multiply it with the thermal velocity, which is $\sim r^{-1/4}$, such that $t_c v_{th} \sim r$ or proportional to the length over which n changes appreciably. For average conditions, the degree of approximation, with which the condition stated at the beginning of this section is fulfilled, is thus for the electrons more or less independent of r (possibly $\sim r^{1/4}$) and similar to conditions found at 1 a. u.

At 10 $R_\odot$ and 4 $R_\odot$ respectively, we have $n = 10^{4.0}$ and $10^{5.1}$ cm^{-3}, $1/\Delta \ln n \approx 10^{11.4}$ cm and $10^{10.8}$ cm, and $T \approx 10^{5/8}$ and $10^{6.0}$. We find $t_c \approx 10^{4.3}$ and $10^{3.5}$ for ions or $10^{2.7}$ sec and $10^{1.9}$ for electrons, which should be halved because of the collisions with ions as before. The bulk velocity should be $\lessapprox 10^{7.1}$ cm/sec and $6 \cdot 10^6$ cm/sec respectively and the resulting time for the ions $\lessapprox 10^{4.3}$ sec and $7.6 \cdot 10^4$ sec, such that now the condition stated above is just fulfilled also for the ions, at least marginally; for the electrons this time interval is now $10^{11.4}$ cm/$10^{8.7}$ cm/sec $\approx 10^{2.7}$ sec, which should be raised to $10^{2.8}$ sec or $10^{2.9}$ sec in order to take spiralling into account. We conclude that for $r \lessapprox 10\ R_\odot$ the macroscopic equations can probably be trusted for the electrons and somewhat less certainly for the ions also. Observational facts bearing on this question will be discussed in section V.

B. <u>The Boltzmann Equation and the Macroscopic Equations for a Fully Ionized Plasma with a Magnetic Field</u>. For the ions (assumed to belong to one chemical species) and the electrons we have separately the familiar Boltzmann equation (written in the same form as in I-9).

$$\frac{\partial f}{\partial t} - \sum w_i \frac{\partial f}{\partial x_i} + \sum \frac{F_i}{m} \frac{\partial f}{\partial w_i} = \left(\frac{\partial f}{\partial t}\right) \text{coll.}$$

(the right-hand-side term is in practice evaluated using the Fokker-Planck collision term. cf. ref. III-2).

Here F includes the electric and magnetic forces $e\underline{E}$ and $e\left[\frac{\underline{v}}{c} \cdot \underline{B}\right]$, (hence $\partial F_i/\partial \omega_i = 0$) and gravity. Since all currents in the plasma are appearing explicitly in the equations, the magnetic induction $\underline{B}$ is everywhere used ($e, \underline{E}$, and $\underline{B}$ in Gaussian units).

We wish to discuss the question whether serious limitations of the macroscopic equations derived from the Boltzmann equation result from the fact that the Lorentz forces are relatively large and that therefore the gyration radii of the charged particles

$$a = \frac{w_\perp}{\omega_c} = \frac{m w_\perp c}{Z_e B}$$

are usually small compared to all other scales of interest, in particular to the mean free path. The gyration radius is e.g. $\approx 10^3$ cm for protons in the corona, or $\approx 10^7$ cm at 1 a.u. ($1/2 \cdot 10^{-4}$ Gauss, $w_\perp$ measured in the co-moving system).

Taking the integral of the Boltzmann equation over velocity space, we obtain after integration by parts of the term containing F_i the continuity equation

$$\frac{\partial n}{\partial t} + \nabla \cdot (n\underline{v}) = \frac{\partial n}{\partial t} + \text{div}\,(n\underline{v}) = 0$$

$$n(r,t) = \iiint f(\underline{r},\underline{w},t)\, dw^3$$

$$\underline{v}(r,t) = \frac{1}{n(\underline{r},t)} \iiint \underline{w} \cdot f(\underline{r},\underline{w},t)\, dw^3$$

The equation of momentum transfer for each species is obtained by taking the integral over velocity space of $m\underline{w}$ times the Boltzmann equation. We obtain at first (cf.[I-9] appendix)

$$\frac{\partial}{\partial t}(nm\underline{v}) + \nabla(n,m\,\overline{\underline{w}\,\underline{w}}) - n\overline{\underline{F}} = \iiint m\underline{w}\left(\frac{\partial f}{\partial t}\right)_{coll} dw^3 .$$

Writing

$$\psi_j = \frac{m}{\Delta v} \sum (\underline{w}_j - \underline{v}_j)(\underline{w}_j - \underline{v}_j)$$

(summation over all particles in the volume element Δv) we arrive at the stress tensor. With

$e(\underline{E} + \frac{1}{c}[\underline{v} \times \underline{B}]) \equiv e\underline{E}^c$ electric force measured in the co-moving system

$-\nabla\varphi \equiv +\underline{g}$ gradient for the gravitational potential

we get finally

$$nm\left(\frac{\partial \underline{v}}{\partial t} + (\underline{v}\,\mathrm{grad})\underline{v}\right) = neE^c - \nabla\psi - nm\nabla\varphi + P_{coll}$$

where P_{coll} is the net gain or loss of momentum due to collisions with particles of the other species (ions or electrons).

The case of strong collisions in a fully ionized gas can be fully handled (that is including transport phenomena) by the Chapman-Enskog technique (cf. Ref. 4). If collisions are rare, such that the Joule dissipation is negligible, a good approximation can be reached if (and only if) the heat flow connected with the third moment of the Boltzmann equation can be neglected. For not too high temperatures--effective wave velocities large compared with the thermal speed of ions or electrons, which is unfortunately not the situation in interplanetary space,--one can thus obtain a closed set of momentum equations (cf. e.g. Bernstein and Trehan[4]). For strong magnetic fields--$a \ll (|\nabla \ln \rho|)^{-1}$, which is easily fulfilled in our case--the pressure tensor becomes essentially diagonal, with two values $p_\perp$ and $p_\parallel$ perpendicular and parallel to $\underline{B}$ respectively. The (adiabatic) invariance of the magnetic moment of a single charged particle gyrating around a line force is expressed by

$$\pi a^2 \frac{e}{c} \frac{\omega c}{2\pi} = \frac{1}{2} \frac{m w^2_\perp}{|B|} = \text{const}$$

and holds for--on the time scale $(\omega_c)^{-1}$ --slow changes of B and E under a wide range of conditions; it leads to a relation for $p_\perp$[1])

$$\frac{d}{dt}\left(\frac{p_\perp}{\rho|B|}\right) = 0$$

with $\frac{m}{k} \cdot \frac{p_\perp}{\rho} = T_\perp$. If the change of $|B|$ is essentially due to changing external conditions, the work done can be related to the induced electric field accompanying the change in $\underline{B}$. In the case of the solar wind, however, a detailed consideration shows that the thermal energy of the degrees of freedom $\underline{B}_\perp$ becomes macroscopic kinetic energy $\parallel \underline{B}$ (cf. e.g. ref. I-9, chapter 1).

Another (longitudinal) invariant, which is usually less well preserved, is related, if operative (that is for vanishing heat flow), to a similar law[+)] for $p_\parallel$ which,

---

+) $d(kT_\parallel B^2/\rho^2)dt = 0.$

with the one given for $p_\perp$, is the analogue to the adiabatic relations of an ordinary gas. A derivation of these equations based on the first law of thermodynamics and on the assumption of zero heat flow has been given by R Lüst[5].

Unfortunately, the condition specified above concerning the heat flow is certainly not fulfilled for the electrons in interplanetary space as may be seen also by using arguments of elementary kinetic theory. Since the relative temperature difference along a mean free path is not really $\ll 1$ --cf. also the measurements of the velocity distribution function of the electrons to be discussed in section V--it is clear that there must be a substantial heat flow (which in fact has been measured recently). Also for the ions, the evidence appears to indicate non-adiabatic conditions (see sections II-C and V; cf. also the model discussed in III-D).

The difficulty of formulating the energy equation in the case of rare collisions is of course related to the fact that in a dilute magnetized plasma energy may be present in a number of different forms and that only its total amount is conserved.

If the stress tensor has effectively only diagonal components and if, furthermore, these are equal (= p), which is the case if the mean free path is small compared to the distance over which the macroscopic quantities n,p,v, change appreciably--e.g. $(\text{grad} \ln n)^{-1}$ --div ψ becomes grad p; we arrive then at the simple form of the equations most often used. Putting

$$\underline{v} \equiv \frac{1}{\rho} (n_i m_i \underline{v}_i + n_e m_e \underline{v}_e)$$ (ρ is mass density; i.e., now referring to ions and electrons respectively)

$$\underline{j} = e(Zn_i\underline{v}_i - n_e\underline{v}_e)$$

assuming--as known from plasma physics--electric neutrality over volumes large compared to the "Debye shielding distance" $\sqrt{kT/4n_e e^2} = 6.9 \cdot \sqrt{T/n_e}$ ($\approx$ 1 cm in the inner corona, $\approx 10^3$ cm near 1 a.u.) and neglecting $\partial \underline{j}/\partial t$ (that is e.g. plasma oscillations), m_e against m_i and quadratic terms in $\underline{j}/en$, we recover first the familiar equation of mass motion for the whole plasma

$$\rho \frac{d\underline{v}}{dt} = \frac{1}{c} [\underline{j}x - \underline{B}] - \nabla p - \rho\nabla\varphi \quad ,$$

and furthermore--after elimination of the terms corresponding to the Hall current--the relation for the current density (Ohm's law)

$$\eta\underline{j} = \underline{E}^c - \frac{m_i}{ez\rho} \nabla p_e \equiv \underline{E}^c + \underline{E}^i \quad ,$$

where now the first term on the right-hand side is the electric field measured in the co-moving system defined by the mass velocity, while the second term (E^i) is equivalent to an "impressed" electric force. This formulation takes into account the particle drifts which, in the case of rare collisions, become important, as does the contribution to $\underline{j}$ of ∇n in the presence of a transverse magnetic field. The resistivity (sec^{+1}) is related to P_{coll} in the momentum equation by

$$\eta \underline{j} e n_e = P_{coll} \quad .$$

The essential point, first demonstrated by Schlüter[6)] and subsequently in more detail by Spitzer (cf. [I-9)]), is that

η is only weakly dependent on the intensity of the magnetic field--more exactly on the ratio of the collision- and the gyro-frequency--such that Ohm's law as stated above holds also for currents perpendicular to B, the resistivity transverse to a strong magnetic field (gyration frequency >> collision frequency) being only approximately doubled. Its numerical value is[I-9], for $\underline{E} \parallel \underline{B}$,

$$\eta = 0.73 \cdot 10^{-8} \frac{\ln \Lambda}{T^{3/2}} \quad \text{e.s.u.}$$

The inclusion of the time derivative of $\underline{j}$ would have brought an additional term to the left-hand side of Ohm's law. Maxwell's equations are now

$$\text{div } \underline{E} = 4\pi e(n_i Z - n_e)$$

$$\text{div } \underline{B} = 0$$

$$\text{rot } \underline{E} = - 1/c \frac{\partial \underline{B}}{\partial t}$$

$$\text{rot } \underline{B} = \frac{4\pi}{c} \underline{j} + \frac{1}{c} \frac{\partial \underline{E}}{\partial t}$$

with

$$\text{div } j = \frac{\partial}{\partial t} (eZn_i - en_e) \quad .$$

It is worth noting that these equations[+)] give a correct description of a physical situation only if all of them (including the equation of mass motion) are taken into account together. Since, in interplanetary space, $p_{\parallel}$ is usually $\neq p$, we note the momentum equation for the direction $\perp B$

+) rot ≡ curl ≡ ∇x

$$\frac{d\underline{V}_{\perp}}{dt} = -\nabla_{\perp}\left(p_{\perp} + \frac{B^2}{8\pi}\right) - \left(\left(\frac{p_{\|} - p_{\perp}}{B^2/4\pi} - 1\right)\underline{B}\,\text{grad}\right)\underline{B}_{\perp}/4\pi$$

These equations may be used to derive the relative separation of charges due to the gravitational field and to the acceleration. From Ohm's law we see that under stationary conditions (constant charge of the sun, $j = 0$) $\underline{E}$ must be equal to $-\underline{E}^i = -(1/eZn_i)$ grad p_e, or, for $Z = 1$ and $p_e = \frac{1}{2}p$,

$$e\underline{E} = -\frac{1}{2}\frac{1}{n_i}\,\text{grad } p = -\frac{1}{2}m_i\left(\underline{g} - \frac{d\underline{v}}{dt}\right)$$

such that, as mentioned before, the electric field compensates the difference of gravitational attraction of the ions and the electrons. Finally the net charge may be obtained from the first equation.

The time derivative of the fourth equation of Maxwell with $B = 0$ is

$$\frac{\partial j}{\partial t} + \frac{1}{4\pi}\frac{\partial^2 E}{\partial t^2} = 0$$

which, substituted into Ohm's law, leads to

$$\frac{-m_e}{4\pi n_e^2}\frac{\partial^2 E}{\partial t^2} + \eta j = E - \frac{1}{eZn_i}\nabla p_e \quad .$$

This equation shows that

$$\omega_p = \left(\frac{4\pi n e^2}{m_e}\right)^{1/2}$$

is a critical frequency for electrostatic plasma waves (for a further discussion of these waves see Ref. I-9). Because of the relation

$$W_p^2 = \frac{kT/m_e}{\lambda_D^2} \quad ,$$

it is the characteristic time scale of the largest statistical fluctuations occurring in the plasma.

C. "Frozen-in" Magnetic Fields and "Force-Free" Magnetic Fields. Since we consider here only slowly varying or constant fields, we neglect $\partial E/\partial t$ and rewrite, using Ohm's law, the left-hand-side of the law of induction integrated over a surface S as follows

$$- c \iint \text{rot } \underline{E} d\underline{S} - \iint \text{rot } [\underline{v} \times B] d\underline{S} =$$

$$- c \oint (\underline{E} + \frac{1}{c} [\underline{v} \times \underline{B}]) d\underline{s} =$$

$$- c \oint (\eta \underline{j} - \underline{E}^i) d\underline{s} \quad .$$

If $\underline{E}^i$ can be neglected, as is easily verified for interplanetary space ($e\underline{E}^i$ is of order $GM_\odot m_i/r^2 << (v/c)eB$), we get

$$- c \iint \text{rot } \underline{E}^c d\underline{S} = - c \oint \eta \underline{j} \, d\underline{s}$$

Here the surface integral, by the law of induction, can be shown (as in the similar case of the theory of vortices) to measure the change of the magnetic flux through a closed line moving with the fluid+); for this the integral over the surface bounded by the moving closed line has to be taken.

+) $\iint \text{rot } [\underline{v} \times \underline{B}] d\underline{S} = \oint B [\underline{v} \times d\underline{s}]$.

The right-hand-side, in astrophysical applications, is usually small in the sense that the time scale of relative changes of the magnetic flux through a surface, or closed line moving with the fluid produced by Ohmic losses, is very long.

$$- c \frac{\iint \text{rot } \underline{E}^c d\underline{S}}{\iint \underline{B} d\underline{S}} = - c \frac{\oint \eta \frac{c}{4\pi} \text{rot } \underline{B} d\underline{S}}{\iint \underline{B} d\underline{S}} =$$

$$- \frac{c^2}{4\pi} \bar{\eta} \frac{\iint \text{rot rot } \underline{B} d\underline{S}}{\iint \underline{B} d\underline{S}} \text{ sec}^{-1} \quad .$$

The resulting time scale is, under normal geometrical conditions, of the order of

$$4\pi \iint dS \div c^2 \bar{\eta} \quad \text{resp.} \iint dS \div c^2 \bar{\eta}$$

(with $\eta \approx \eta_\perp$ in typical situations, such as magnetic filaments or a conical sector in interplanetary space).

In the corona, for instance, η would be of the order of 10^{-16} sec, and 1000 km would be the diameter of the finest structures; the resulting time scale is 10^{12} sec or some 10,000 years. In interplanetary space, structures would usually be much larger, while η near 1 a.u. is probably $\lesssim 10^{-14}$ sec. The resulting time scales are in any case very much longer than those of changes in the corona ($\lesssim 1^d$) or in interplanetary space.

One may hence regard the magnetic lines of force as frozen into e.g. the solar wind plasma, which carries them far into interplanetary space. As we shall see later (section IV-A), this consideration leads at once to the now familiar picture of the spiral pattern of the lines of force in interplanetary space which is typical for average quiet conditions.

Another relation, which is derived easily for frozen-in fields, is

$$\frac{d}{dt}\left(\frac{\underline{B}}{\rho}\right) = \left(\frac{\underline{B}}{\rho}\,\mathrm{grad}\right)\underline{v} \quad .$$

The (normally justified) assumption $\eta\,\underline{j} = 0$, that is $\underline{E}^i = 0$ (though $\underline{E}$ may be large) together with

$$\iint \mathrm{rot}\ \underline{E}^c d\underline{S} = 0$$

implies that, in a co-moving system of reference, only the small electric fields, which may arise from charge separation in the way discussed already, will normally occur. Adjacent bundles of magnetic lines of force of different direction which are often observed in interplanetary space, are of course separated by sheet currents. Another consequence, which is intuitively obvious from the concept of "frozen-in" lines of force, though it was not recognized until Alfvén's pioneering work of 1942[7], is the possibility of the appearances of new types of waves, "magnetohydrodynamic waves", with a characteristic velocity of

$$v_A = \frac{|B|}{\sqrt{r\Pi\rho}} \quad .$$

If collisions are rare, acoustic waves are maintained by the electric fields associated with the relative motion of the ions and the electrons for wave frequencies below the ions' plasma frequency. The sound velocity of these waves is given by the relation

$$V_s^2 = j \frac{k(T_i + T_e)}{m_i}$$

with $j = 3$ in the absence of collisions. The main damping mechanism of the waves is a microscopic effect, the so-called Landau damping, which appears to play an important role in the solar corona[8].

Microscopic wave-particle interactions are generally held responsible for the observed (but theoretically not yet fully understood) fact that, in a dilute magnetized plasma streaming with hypersonic velocity around an obstacle like the geomagnetic field, collisionless shock transitions appear in approximately that position which follows from ordinary gas dynamics for the given velocity field. The fine structure of the bow shock upstream from the earth appears to reveal the hybrid electron-ion gyration frequency $eB/c\sqrt{m_e m_p}$.

We note finally that, in suitably twisted magnetic fields, rot $\underline{B}$ can be approximately or even exactly $\| B$ over large volumes, such that the magnetic volume force $[\text{rot } \underline{B} \times \underline{B}]$ vanishes there. Such "force-free" magnetic fields play an important role in astrophysics, for instance in solar physics. In the solar corona, $\nabla(B^2/4\pi)$ is above strong surface fields, usually much larger than the product of the density and the local value of gravity (and acceleration) and than ∇p; also the magnetic volume force must therefore be $\ll \nabla(B^2/\pi)$.

A hydrodynamic analogon to the force-free magnetic fields are the "Beltrami flows" of hydrodynamics (with $\underline{v} \|$ rot $\underline{v}$).

D. <u>A Solar Wind Model, Based on the Vlasov Equation</u>. Since the violation of the conditions which must be fulfilled if the macroscopic equations should legitimately be used, and which were discussed in III-A, is particularly serious for the ions, a model of the solar wind was investigated by Jockers[9], in which the Vlasov equation was used for describing the ions for $r > r_o = 2.5\ R_\odot$ or $4.5\ R_\odot$, while for the electrons the usual macroscopic equations of momentum were taken to hold. In order to make the problem tractable, some rather severe simplifications were introduced: (1) The ions were assumed to move only radially. (2) As a boundary condition, a one-dimensional Maxwellian velocity distribution was assumed for the ions at r_o, while collisions were entirely disregarded outside r_o. (3) Instead of the energy equation, an adiabatic relation between temperature and density was assumed for the electrons. (4) Magnetic fields were neglected; since a radial field does not affect ions moving radially, this neglect should not be too serious for $r \lesssim 100\ R_\odot$.

Under these circumstances, the total energy E of each ion, ($V(r)$ is the electric field, which ensures that the ions and the electrons move together; m_i and v are mass and velocity of an ion, G is gravitational constant, $M_\odot$ is mass of the sun; the total potential F is defined by the last equation)

$$E = \frac{1}{2} m_i v^2 + G \frac{M_\odot}{r_o} m_i \left(1 - \frac{r_o}{r}\right) - eV(r) =$$

$$\frac{1}{2} m_i v^2 + F(r),$$

is constant and the Vlasov equation can be written down e.g. as the continuity equation of the ions of the energy $E \pm (1/2)\Delta E$. The condition for a stationary state if ($f_i(E,r)$ is the velocity distribution function)

$$0 = \frac{1}{r^2} \frac{\partial}{\partial r} (r^2 v f_i) = \frac{1}{r^2} \frac{\partial}{\partial r} \left(r^2 f_i \sqrt{\frac{2}{m_i} (E - F(r))}\ \right).$$

The solution depends on the number of extremal points of the potential function $F(r)$.

Anticipating the results we note that the total potential energy $F(r)$ has a maximum F_m at some distance $r_m > r_o$ (several times r_o in typical cases), beyond which it declines, for the solutions of interest, only slowly; the potential of the accelerating electric field rises monotonically outward. Particles with $E < F_m$ cannot move out to large distances, as is the case in every theory of evaporating atmospheres.

The distribution function now becomes

$$f_i(r,E) = \begin{cases} 0 & \text{for } E \le F(r) \\ \dfrac{r_o^2}{r^2}\dfrac{n_o}{\sqrt{\pi k T_o}}\dfrac{e^{-\frac{E}{kT_o}}}{\sqrt{E - F(r)}} & \text{for } r_o \le r < r_m \\ 0 & \text{for } r_m \le r \end{cases} \Bigg\} F(r) < E < F_m$$

$$\qquad \dfrac{r_o^2}{2r^2}\dfrac{n_o}{\sqrt{\pi k T_o}}\dfrac{e^{-\frac{E}{kT_o}}}{\sqrt{E - F(r)}} \quad \text{for} \quad F_m \le E$$

in which $F(r)$ is the unknown function and r_m is an unknown parameter. The gradient of the electric potential being essentially given by $-(1/n)\nabla p_e$ on one hand and equal to $m_i(v_k\nabla v_k + g(r))$ on the other hand for each group of ions specified by its energy $E_k v$ (g is gravitational acceleration) it is easy to see that the equations of continuity and of momentum together with the condition of quasineutrality and a relation specifying T_e either as a function of n_e or of position, do indeed, with proper boundary conditions at $r = r_o$, determine the solution. The actual procedure, however, turns out to be too complicated to be described here in detail, for instance by the fact, mentioned already, that for $r < r_m$

there are two classes of ions, one of which falls back to the sun. Furthermore a critical radius $r_c > r_m$ appears, where the bulk velocity becomes equal to the velocity of certain waves, as in Parker's theory. The condition that the solution should be well behaved determines the final choice of r_m. In the collision free region $r > r_o$, a local ion "temperature" can be defined by

$$kT_i(r) = \frac{m_i}{n_i} \int (v - u_i)^2 f_i dE$$

where u_i is the mass velocity of the ions which, because of charge neutrality, must be equal to that of the electrons. For T_i, the asymptotic solution for large r and $|F| \gg kT$,

$$T_i = T_o \cdot \frac{kT_o/2}{F_m - F(r)} \left(1 + 3 \frac{kT_o}{F_m - F(r)} + \dots \right)$$

is obtained, which for especially simple conditions can also be derived directly. F varies only slowly with r at large distances, and, in consequence, also u_i. Also for u_i an asymptotic expression is found:

$$u_i = \sqrt{\frac{2}{m} (F_m - F)} \left(1 + \frac{1}{2} \frac{kT_o}{F_m - F} - \frac{1}{2} \frac{k^2 T_o^2}{(F_m - F)^2} + \dots \right).$$

The adiabatic exponent for the electron component was chosen as 1.0 (isothermal case), 1.1 (probably the best choice of the three), and 1.2. The following two models, the second of which is probably closer to the average conditions, were considered in more detail:

$$\left.\begin{array}{ll} \text{(I)} & r_o = 2.5\ R_\odot,\ n_{e,o} = 7 \cdot 10^5\ \text{cm}^{-3} \\ \text{(II)} & r_o = 4.5\ R_\odot,\ n_{e,o} = 10^5\ \text{cm}^{-3} \end{array}\right\} \quad T_o = 1.026 \cdot 10^6 .$$

The results for $r = 215\ R_\odot \approx 1$ a.u. (the values in brackets for n_e hold for 30 $R_\odot$) are given in Table II.

The agreement with the actual values of n, u_i and T_i is probably as good as should be expected in view of the simplifications used (see especially the ratio of the densities at the two distances 30 and 215 $R_\odot$). A point of particular interest, however, is the fact that at a large distance the maximum of f in velocity space does not coincide with the bulk velocity; the velocity dispersion in the antisolar direction is larger than in the opposite one because a certain not insignificant fraction of the slower than average ions has fallen back to the sun. This is qualitatively what has been observed (see section V-A).

SOLAR WIND CALCULATIONS, BASED ON THE VLASOV EQUATION

Table II

	$u_i \cdot 10^{-7}\ \frac{km}{sec}$			$T_i \cdot 10^{-5}\ {}^oK$			$T_e \cdot 0^{-5}\ {}^oK$			$\log(n_e[cm^{-3}])$		
α =	1	1.1	1.2	1	1.1	1.2	1	1.1	1.2	1	1.1	1.2
I	3.22	1.70	.81	.65	1.63	3.46	10	2.79	.64	.38	.19	- .18
				(1.39	2.88	5.30)				(2.28	2.12	1.80)
II	3.23	2.10	1.27	.65	1.23	2.27	10	3.81	1.41	.65	.69	.69
				(1.38	2.19	3.52)				(2.58	2.62	2.62)

References of Section III

1) G. Chew, N. Goldberg, F. Low, Proc. Roy. Soc. 236A, 112 (1956).

2) W. B. Thompson, An Introduction to Plasma Physics, Pergamon Press, Oxford (1962).

3) J. Wilcox, University of California, Berkeley, Techn. Rep. Ser. 9, January 1968.

4) I. Bernstein, S. Trehan, Nucl. Fusion 1, 3 (1960).

5) R. Lüst, Introduction to Solar Terrestrial Relations, Chapter 12, Reidel Publish. Company, Dordrecht, (1965).

6) A. Schlüter, Z. Naturforschg. 5a, 72 (1950).

7) H. Alfven, C. Fälthammar, Cosmical Electrodynamics.

8) N. d'Angelo, ESRIN Preprint (1968). C. Parkinson, K. Schindler, ESRIN Preprint (1968).

9) K. Jockers, Mitt. Astron. Ges. 24, 187 (1968).

IV. THEORY OF THE INTERPLANETARY MAGNETIC FIELDS AND THE TRANSPORT OF ROTATIONAL MOMENTUM

A. The Large Scale Structure of the Magnetic Field. The concept that lines of force of the extended surface fields of the sun are convected+) into interplanetary space by the solar wind, in conjunction with that of frozen-in magnetic fields (section III-C), was used by Parker already in 1958 (Ref.I-5) to predict the spiral structure of the interplanetary magnetic field, which was indeed subsequently observed to be its prevailing large scale property (Section I-C). The argument can also be expressed by stating that the "Chapman spiral", known from the theory of recurrent geomagnetic disturbances, should not only be the locus of a certain bundle of magnetic lines of force originating in the solar "M region", the source for a series of recurrent disturbances, but also the typical interplanetary magnetic field configuration for quiet periods of time. The dependence of the bulk velocity on solar distance fixes the shape of the spiral as soon as the extent of co-rotation of the solar corona--more specifically the angular velocity as a function of distance from the sun--is known; this can be found only by considering the momentum balance in the azimuthal direction and the transport of rotational momentum. The time lag of 4 1/2 days, defined by the correlation of the magnetic polarities on the sun's surface and near 1 a.u., together with the bulk velocity directly observed, shows, however, that the radius of the co-rotating region around the sun must be a rather small fraction of 1 a.u. That only the lines of force of extended weak magnetic regions are carried outward can be understood by comparing the magnetic tensions involved with the pressure forces: A field strength

---

+) For a recent discussion of the concept of the motion of field lines see D. P. Stern[1].

of, say, 2 Gauss leads to a magnetic tension of 0.16 dyn/cm^2; since there is every reason to think that the solar fields become effectively almost force-free in the corona[+)], the actual volume forces present should be only a fraction of this amount, which is then of the same order as the pressure forces. Still it is obvious from solar observations that the much stronger magnetic fields of sunspots and around coronal condensations are essentially confined to the corona itself.

Before considering the balance of momentum in more detail, we note that H. Babcock[2)], in his model of the periodic changes of the solar magnetic fields, assumed that there is at times in the outer atmosphere of the sun also some severing and reconnecting of magnetic lines of force. This would mean that bundles of closed lines of force, with a sizeable total magnetic flux, are then carried by the solar wind through interplanetary space. This possibility was further discussed by L. Biermann and R. Lüst[I-24)]; recently C. Winge, P. Coleman, L. Davis, Jr., E. Smith and D. Jones[3)] found some observational evidence in the Mariner IV data that such events may indeed occur.

B. The Angular Momentum of the Solar Wind. In this section we follow mainly the work of Modisette[4)] and of Weber and Davis[5)] (notation essentially as in [5)]). Spherical polar coordinates r, ν, φ are used with $\partial/\partial\varphi = 0$, because symmetry around the axis is assumed; since attention is given mainly to the vicinity of the equatorial plane, B_ν is assumed to vanish. The sign of the sun's rotation and of the partial

[+)] For an early treatment of the transport of angular momentum by force-free magnetic fields see R. Lust and A. Schluter[9)].

co-rotation of the interplanetary plasma is positive (right-handed system). The continuity equation reads $(u = v_r)$

$$\rho u r^2 = m \approx 10^{11.0} \text{ g/sec,ster}$$

$\underline{E}^c = 0$ (that is $\underline{E} = - [(\underline{v}/c) \times \underline{B}]$, cf. Section III-C) and rot $\underline{E} = 0$ (steady state) yield for the φ component of the latter

$$\frac{1}{r}\frac{d}{dr}[r(uB_\varphi - v_\varphi B_r)] = 0$$

$$r[uB_\varphi - v_\varphi B_r] = \text{const} = - \Omega R_\odot^2 B_{r,o}$$

$$= - \Omega r^2 B_r$$

$((uB_\varphi)_o \ll (v_\varphi B_r)_o)$ because of $u_o \ll v_{\varphi,o}$ and $(B_\varphi/B_r)_o$ is finite owing to the constancy of the magnetic flux $(\Omega R_o = 2.00$ km/sec). The φ component of the equation of motion is (viscosity and other effects contributing to the flow of angular momentum of rotation being neglected; see the additional note below),

$$\rho \frac{u}{r}\frac{d}{dr}(rv_\varphi) = \frac{1}{4\pi}[(\text{rot } B) \times B] =$$

$$\frac{B_r}{4\pi r}\frac{d}{dr}(rB_\varphi) \quad .$$

Furthermore, we have

$$B_r \cdot \frac{B_r}{4\pi\rho u} = \frac{r^2 B_r}{4\pi m} = \text{const}$$

by which we get

$$mrv_\varphi - \frac{1}{4\pi}(r^2B_r)(rB_\varphi) = mL \text{ (constant of integration)}$$

With $|B_r| = |B_\varphi| = 5_\gamma (B_rB_\varphi < 0)$ and (tentatively) $v_\varphi \approx 9$ km/sec (cf. section I-C) the two terms are approximately in the ratio 2 : 1 near 1 a.u., their sum being $mL \approx 2.0 \cdot 10^{30}$ dyn,cm/ster.

In order to derive $v_\varphi(r)$ we define a Mach number M_A by

$$M_A^2 = \frac{\rho u^2}{B_r^2/4\pi} = \frac{mur^2}{(r^2B_r)^2/4\pi} = \frac{4\pi m^2}{(r^2B_r)^2} \cdot \frac{1}{\rho} \quad .$$

This gives

$$\frac{v_p}{r\Omega} = \frac{M_A^2 Lr^{-2}\Omega^{-1} - 1}{M_A^2 - 1}$$

M_A^2 being $\approx 10^{2.0}$ near 1 a.u., but < 1 for $r \gtrsim R_o$, it is clear that it must pass through a region with $M_A = 1$; regularity of the function $v(r)$ requires

$$L = \Omega r_a^2$$

where the index a refers to the critical "Alfven" point with $M_A = 1$. From this, r_a^2 may be derived, but independently also by the equation defining M_A^2 which shows that

$$M_A^2(r) = \frac{\rho a}{\rho(r)}$$

and

$$v_\varphi = \frac{\Omega r}{1 - M_A^2} \cdot \frac{u_a - u}{u_a}$$

The azimuthal velocity $v_\varphi(r)$ is thus related to $u(r)$, that is $\rho(r)$, and also to $M_A(r)$[+)].

Finally B_φ is given by

$$\frac{B_\varphi}{B_r} = - \frac{\Omega r}{u_a} \cdot \frac{1}{1 - M_A^2} \cdot \frac{r_a^2 - r^2}{r_a^2}$$

which for $r^2 \gg r_a^2$ and $M_A^2 \gg 1$ becomes the spiral

$$\frac{B_\varphi}{B_r} = - \frac{\Omega r}{u_a} \cdot \frac{\rho}{\rho_a} \cdot \frac{r^2}{r_a^2} = - \frac{\Omega r}{u}$$

$$r B_\varphi = - \Omega(r^2 B_r)/u$$

whereas for $r^2 \ll r_a^2$ we have simply $B_\varphi/B_r = - r\Omega/u_a \lessapprox 10^{-1}$, by order of magnitude for $r \gtrapprox R_\odot$. This shows that there the first term in the integral of the momentum equation is small compared to the second term.

Weber and Davis proceed to derive $v(r)$ by means of the equation pertaining to the radial momentum balance and the energy equation derived from it. We note two points. The radial component of the magnetic volume force is

$$\frac{1}{4\pi} [\text{rot } \underline{B} \times \underline{B}]_r = - \frac{1}{4\pi r} \frac{\partial}{\partial r} (r B_\varphi) \cdot B_\varphi$$

+) The value of v_φ near 1 a.u. derived from this last equation, $(u - u_a)/u_a$ being $\lessapprox 1$ for $r_a \approx 30\ R_\odot$ (cf. Table 1 section I) is less than half the observed value. This may be due to the incompleteness of the theory (see additional note) and/or to the fact that the observed values used do not refer to the same periods of time.

$$= \frac{-1}{8\pi r^2} \cdot \frac{\partial}{\partial r} (rB_\varphi)^2$$

with

$$(rB_\varphi) = \Omega \frac{(r^2 B_r)}{r_a^2 u_a} \frac{r^2 - r_a^2}{M_A^2 - 1}$$

which is seen to be approximately independent of r near 1 a.u., such that its contribution to the radial momentum balance is small against $\nabla(B^2/4\pi)$ or ∇p.

From this the sum of the contributions of the Poynting vector and of v_φ^2 to the energy flux is obtained in the form

$$F = m \frac{\Omega^2 r_a^4}{2r^2} \left[1 + \frac{(2M_A^2 - 1)(r^2 - r_a^2)^2}{r_a^4 (M_A^2 - 1)^2} \right]$$

$$= m \left[\frac{1}{2} v_\varphi^2 - \frac{B_\varphi B_r}{4\pi\rho} \frac{\Omega r}{u} \right]$$

$$= m \left[\frac{1}{2} v_\varphi^2 - \Omega \frac{rB_\varphi (r^2 B_r)}{4\pi m} \right] .$$

Its order of magnitude near 1 a.u. is 2.10^{13} (erg/g) · m (or ≈ 2% of the flux of kinetic energy, the contribution of the term containing v_φ being insignificant), and is seen to be approximately constant for larger values of r. The second point refers to the character of the singularity at r_a. This turns out to be more complicated than that of the one previously discussed in section II-A, essentially because there are several characteristic velocities of magnetohydrodynamic waves, the ratio of which to the bulk velocity affects the balance of momentum (cf. the equation defining M_A, and sections II and III above).

Finally an approximate solution is obtained, which qualitatively satisfies the given boundary conditions at the sun's surface and near 1 a. u. That the quantitative agreement turns out to be not really satisfactory is ascribed by the authors mainly to the difficulties concerning the energy equation (see sections II and III)[+].

The value of the torque acting on the sun, thus found, is large enough to be of considerable influence on its evolution. The time scale of change, by a factor e, of the sun's present angular momentum of rotation is found by Weber and Davis and by Modisette to be several 10^9 years, which is comparable to its age. It is conceivable that in earlier more active stages of the sun's evolution the rate of loss of angular momentum was larger.

Additional Note: In three papers by Yen[6)], by Schubert and Coleman[7)], and by Meyer and Pfirsch[8)], the azimuthal equation of momentum and its integral expressing the flux of angular momentum have been modified to include effects neglected in the papers discussed so far.[4), 5)] Yen took into account a finite (constant) value of the electric resistivity, whereby it becomes possible to obtain equations which are regular at the critical layer and to apply boundary conditions with an arbitrary value of the azimuthal component of the magnetic field at the sun's surface. Schubert and Coleman discussed the influence of fluctuations and plasma waves on the state of rotation of the interplanetary plasma; they suggest these effects might be quite considerable. Meyer and Pfirsch showed that the (observed) anisotropy of the

----

+) As far as the viscosity is concerned, it can be shown that its influence in the azimuthal momentum balance is near $r = R_\odot$ of the same order as that of the other terms, although its influence in the integral at large distance should be much smaller.

velocity distribution of the ions (see section V-A below) affects significantly the flux of angular momentum. In this case the position of the critical layer is defined by the velocity of torsional waves ($V_r^2 = (B_r^2/4\pi\rho)(1 - p_{\parallel} - p_{\perp})4\pi/B^2$); if the anisotropy is close to the limit set by the onset of instabilities (see section V-A below), the term mrv_{φ} represents the total flux of angular momentum.

References of Section IV

1) D. P. Stern, Space Sci. Rev. 6, 147 (1966).

2) H. W. Babcock, Ap. J. 133, 572 (1961).

3) C. Winge, P. Coleman, L. Davis, E. Smith, D. Jones, Publ. No. 651, p. 13, Inst. of Geophys. and Planet. Physics, UCLA (1968).

4) J. L. Modisette, J. Geophys. Res. 72, 1521 (1967).

5) E. J. Weber, L. Davis, Jr., Ap. J. 148, 217 (1967).

6) K. T. Yen, AIAA Paper No. 68-690, June 1968.

7) S. Schubert, P. Coleman, Publ. No. 651, p. 12, Inst. of Geophys. and Planet. Physics, UCLA (1968).

8) F. Meyer, D. Pfirsch, contribution to the Kleinheubach meeting 1968 of the German section of the U.R.S.I. (to appear).

9) R. Lüst, A.Schlüter, Z. Astrophys. 38, 190 (1955).

General: E. Parker, reference I-5.

A. Dessler, reference I-20.

R. Lüst, reference I-7.

V. MICROSCOPIC AND SMALL SCALE PHENOMENA

A. Observations. The most important microscopic feature observed in interplanetary space pertains to the velocity distribution functions of the ions and the electrons. That of the ions (see, for instance, Fig. 12 in Axford, reference I-6) deviates in two respects from Maxwellian distribution. First of all it is fairly anisotropic, the line of maximum deviation from symmetry being parallel to the magnetic field vector. This holds, not only statistically, but also for individual measurements (see Axford, Fig. 13 in reference I-6). The thermal motions parallel to the magnetic field exceed, therefore, those perpendicular to it, normally in the ratio $\sqrt{2}$: 1, but often by a factor of $\sqrt{5}$ up to $\sqrt{10}$. Second, the anisotropy is of such a kind that the velocity distribution function is considerably more extended away from the sun than in the direction of the sun. As a consequence, the maximum of the velocity distribution function does not at all coincide with the average velocity, and one value of $T_{\parallel}$ is not sufficient to describe the physical situation even roughly.

The velocity distribution function of the electrons shows qualitatively the same deviations from isotropy, although these are much less pronounced (of the order of 10%), such that they can just be measured. The first such measurements have just been reported by Montgomery, Bame, and Hundhausen[1)].

The direction of the average velocity shows, in addition to the azimuthal component of around 10 km/sec mentioned already, fluctuations in relatively short periods of time, for instance, by several degrees of arc in approximately five minutes, which, combined with an average velocity of 400 km/sec, correspond to a wave length of only 120,000 km, or less than 1/1000 of an astronomical unit.

The smallest features in the density distribution are the fluctuations of a diameter of 100 or some 100 km observed by radio-astronomical techniques. Still smaller structures are seen in the bow shock of the earth, the fine structure of which reveals substantial changes over distances of the order of 10 km.

Filamentary structures of a million kilometer diameter by order of magnitude have often been observed from space vehicles. Discontinuities of different types--with or without pressure jumps balanced by a jump in the magnetic field strength--seem to be regular features in the interplanetary plasma[2),5)]. An often observed special case are the tangential discontinuities, in which there is neither mass flow nor magnetic flux through the discontinuity. The plasma shear velocity is often several times 10 km/sec, that is of the order of the proton thermal velocities, whereas the changes of the magnetic field direction may be quite substantial. In the cases first mentioned, the magnetic field jumps indicate those of the total pressure, while the ion pressure is often independently known from the plasma measurements. In such cases, the comparison[2)] shows that often it is essentially the electron pressure which balances the magnetic pressure, with electron temperatures perpendicular to $\underline{B}$ of the order of 10^5 K.

B. Theoretical. The asymmetry of the velocity distribution function is usually described by indicating two different temperatures: one ($T_{\|}$) pertaining to the direction parallel to the magnetic field vector, the other ($T_{\perp}$) to the plane perpendicular to it. The asymmetry of the velocity distribution generally found for the charged particles in the sense $T_{\|} > T_{\perp}$

is readily explained from the invariance of the magnetic moment. That it is more pronounced for the protons than for the electrons is to be expected from the considerations given in III-A. Microscopic instabilities[+)] which should prevent the asymmetry from becoming too large (say when $p_{\parallel} - p_{\perp} > (B^2/4\pi)$ or $p_{\parallel} > 2p_{\perp}$) are the "firehose" and the Alfvén wave instabilities, which have the largest growth rate for $\lambda \approx a_p$ (gyration radius of protons); this might explain the appearance of the density fluctuations with dimensions of the order a_p mentioned in V-A. We note, furthermore, that the "temperatures" should really be reckoned from the maximum velocity distribution functions, not from the bulk velocity, because the rear part of the velocity distribution is, so to speak, missing. This asymmetry with respect to a plane $\perp \underline{B}$, which effectively means that the third moment (and the heat flux) should not be neglected without further discussion, is apparently explained for the first time by the work of Jockers (Section III-D).

The fact that the difference of $T_{\parallel}$ and $T_{\perp}$ is regularly seen for the ions appears to indicate that the last interaction affecting the velocity of an ion with respect to the magnetic field vector has taken place much nearer to the sun, at r_o, say r_o being a fraction of an astronomical unit which must be fairly small in those cases where the

---

+) cf. Parker[I-5] or T. Stix[V-3]. Observational evidence that the instability criterion $p_{\parallel} - p_{\perp} > B^2/4\pi$ is really operative in interplanetary space, has been discussed recently by N. Kawashima, GSFC publication X-616-68-271. This criterium is related to the expression for the velocity of torsional waves given in IV, additional note, as is easily verified, and to the expressions for the momentum equation $\perp \underline{B}$ in section III-B.

asymmetry is most pronounced. Thus from the dependence of $|B|$ on solar distance, an assumed reduction of $T_\perp$ by a factor 2 or 5, with constant magnetic moment, would indicate $r_o \approx 0.6$ or 0.4 a.u. respectively as discussed in IV-2. In comparing these figures with the observed ones, it is important to take account of the fact that most of the energy lost from $T_\perp$ is not increasing $T_\parallel$ (cf. III-B) and that $T_\parallel$ decreases also, though probably much more slowly (cf. III-D). Since, as we shall see below, the microscopic velocity distribution is not only affected by collisions in the sense of kinetic theory, but also by the interaction with waves, the distance measured from the earth will ordinarily be smaller than would be estimated from kinetic theory. In fact, we found in Section III-A that the transition to almost collision-free conditions should take place under normal flux conditions at a solar distance of the order of only 10 solar radii or less. That the difference of $T_\perp$ and $T_\parallel$ is normally considerably smaller for electrons is also precisely in line with the estimates given in III-A.

Concerning the equilibrium conditions near discontinuities, we note as a special case of the general equation of motion, with anisotropic pressures (curvature of $\underline{B}$ being neglected),

$$\frac{B^2}{8\pi} + n_p kT_{p\perp} + n_\alpha kT_{\alpha\perp} + (n_p + 2n_\alpha)kT_{e\perp} = \text{const}$$

across the discontinuity, which has been found to be consistent with the observations on many occasions (n_α and T_α have been written separately, since there is some evidence that T_α is several times T_p); the suffix α refers to α particles.

In conclusion, we make some remarks on ion acoustic waves and the damping mechanism discovered by Landau already in 1946 (see references I-9, L. Spitzer, and V-3, T. Stix). From Ohm's law, as formulated near the end of Section III-B, it can be concluded that, for slow waves, the left-hand side of Ohm's law, as formulated there vanishes: the reason is that the resistivity is low and that, for wave frequencies which are small compared to the plasma frequency, also the other term on the left-hand side of the equation referred to is insignificant. To be specific, it should be mentioned that the wave lengths which should be considered are from 100,000 km upwards. In other words, for such a wave, the local electric field intensity is just balanced by the impressed electric force which is produced by the gradient of the electron pressure connected with the wave in question. If we now look at such a travelling longitudinal wave in a system moving with it, we have a distribution of the electric potential which is independent of time as long as no damping occurs. It can then be shown that particles exist which will be trapped in a minimum of the electric potential. Such a particle will therefore be reflected over and over again; and, by this process, there will be an interaction between these particles and the wave with the result that the wave may become gradually damped. The time scale of the Landau damping of such waves has been considered by d'Angelo for the solar corona (1968, unpublished) and recently by Pfirsch and myself for interplanetary space (unpublished so far). The result is that, as far as interplanetary space is concerned, the damping may occur over a distance L parallel to $\underline{B}$ given by the approximate relation

$$L = \frac{1}{k_{\|}} \frac{V}{v_e} \frac{m_i}{m_e} = \frac{\lambda}{2} \frac{V}{v_e} \frac{m_i}{m_e} \approx 40\lambda$$

($\lambda, k_{\parallel}$ are wavelength and wavenumber, v_e is thermal velocity of the electrons, V is bulk velocity). $L \lesssim 10^{14}$ cm $\approx$ 7 a.u. for instance would thus demand $\lambda \lesssim 0.2$ a.u. (that is $\nu^{-1} \approx$ 1 day), such that many of the observed waves should contribute to the heating of the ions. The energy of such waves is, of course, drawn from the inhomogeneities of the velocity field and of the density distribution+), which are known to be quite pronounced near 1 a.u.

+) cf. Reference I-20, Chapters 5 and 6.

References of Section V

1) M. D. Montgomery, S. J. Bame, A. Hundhausen, J. Geophys. Res. 73, 4999 (1968).

2) L. F. Burlaga, Solar Physics 4, 67 (1968).

3) T. H. Stix, The Theory of Plasma Waves, McGraw Hill Book Company, Inc. (1962).

4) L. Landau and E. Lifshitz, Electrodynamics of Continuous Media.

5) J. L. Siscoe, et al., J. Geophys. Res. 73, 61 (1968).

VI. THE TERMINATION OF THE SOLAR WIND

A. Conditions in Interstellar Space. Near hot, luminous stars, the hydrogen, the main constituent of the interstellar gas, is completely ionized ("H II region") out to a distance of some parsec (1 pc = $2.06 \cdot 10^5$ a.u.) or more. Of the other atoms, carbon and various metals give rise to a degree of ionization of the order of 0.1% if the hydrogen is effectively non-ionized ("H I region").

Since no such star is sufficiently close to the sun (the nearest being at distances of several ten pc), and no other evidence such as surface emission in the hydrogen line $H\alpha$ (arising from recombination) indicates ionized hydrogen near the solar system, it is generally assumed that the interstellar hydrogen around the solar system is not or is only weakly ionized.

It should be noted that, owing to the absorption of the stellar Lyman continuum quanta by interstellar atomic hydrogen, the boundaries between H II and H I regions must be rather sharp, as was found first by Stromgren in 1938. The density in H I regions is ordinarily in the range $\lesssim 0.1$ to 10 atoms/cc, the higher values pertaining to condensations or so-called clouds. The average density in our part of the galactic system is somewhat below 1 cm^{-3}, probably ≈ 0.7 cm^{-3}. Our "vicinity", in the sense of the next few ten pc, is likely, however, to be in the low density range. All UV spectra of stars available so far, showing the interstellar Lyman α (1216 Å) line in absorption, indicate, in the directions in question, only $\lesssim 0.1$ H atoms/cc. We shall hence work with the value 0.25 atoms/cc, or $0.5 \cdot 10^{-24}$ g/cm^3

(close to the value 0.2 cm^{-3} recently proposed by Heiles[10] for intercloud regions in an analysis of high resolution radio-astronomical observations in the H I 21 cm line).

The temperature in H I regions being most probably between 10^2 and 10^3, the ordinary gas pressure is only of the order of 10^{-14} dyn/cm^2. There are, however, turbulent motions with approximately $10^{11.4}$ erg/g ($\gtrsim 10^{-13}$ erg/cm^3) and, in the direction of the apex of the solar motion, with approximately 20 km/sec with respect to the local rest frame, a momentum flux from that direction of $2 \cdot 10^{-12}$ dyn/cm^2. Of the same order is the local energy density of the interstellar magnetic fields and that of the cosmic high energy particles ("cosmic rays"). The interaction between the non-ionized gas, the (ionized) plasma, the magnetic field, and the cosmic rays is strong enough to couple them effectively together over distances of interest in the present connection.

B. The Outer Shock Front Defining the Heliosphere. The term heliosphere is used here to indicate that region around the sun which is pervaded by a hypersonic solar wind (cf. Hundhausen, 1968[7]). The momentum flux of the solar wind under average conditions is (m defined in II-B) Equation 1.

$$10^{7.6} \cdot m\, r^{-2} \; g/cm^{-1}\, sec^{-2}$$

or

$$\approx 2 \cdot 10^{-8}\, (r')^{-2} \; dyn/cm^2$$

where r' is measured in a.u.'s. The existence of a cavity in the interstellar medium produced by the solar wind was first pointed out by L. Davis[1], and some years later F. Clauser[2] noted that a shock front should be present at about that distance where the momentum flux of the solar wind becomes smaller than the pressure of the interstellar magnetic fields $B^2/8\pi$ or the sum of all the pressures which would interact with the solar wind (see subsection a).

From this argument, the shock front would be at a distance of the order of 100 a.u. or somewhat less (≈ 35 a.u.), if the intensity of the interstellar magnetic field in our neighborhood would be as high as $2 \cdot 10^{-5}$. If, however, not the whole momentum flux of the solar wind interacts with the interstellar pressure at the shock front, as should be the case if interstellar neutral atoms penetrate the shock front in large numbers and interact with the solar wind already inside the heliosphere, then the momentum balance would be affected in the sense that the shock front would be closer to the sun by at least the root of the reduction factor. In that case the whole momentum balance would be changed in a way which does not seem to have been discussed in the literature other than in relation to comets for which similar questions arise[9)]. The qualitative net result should be that the shock front should move still closer to the sun; but, in the absence of detailed calculations, it is impossible to say by how much.

Outside the shock front, a temperature of the order of 10^7 °K should be reached by the thermalization of the energy of mass motion. Since the pressure can be only of the order given by the interstellar conditions, the density must be quite low, of the order of 10^{-3} ions/cc.

It should be noted that considerations largely based on the observed Lyman α intensity outside the geocorona have led Patterson, Johnson, and Hanson, and more recently Hundhausen, to use these data to derive the position of the shock front. It must be remarked, however, that the question of how the Lyman α intensity of the sky background is produced appears to be quite complex and that new observational data might well modify this interpretation of the Lyman α emission of the sky.

C. <u>The Transition Region Outside the Heliosphere</u>. Concerning the state of ionization, we observe first that the solar Lyman continuum (Lc) quanta ($\lambda < 911$ Å) are absorbed by atomic hydrogen with a cross section of roughly $1/2 \cdot 10^{-17}$ cm^{-2}, that is to say, by a column of some 10^{17} atoms/cm^2. As a consequence, the main absorption of these quanta takes place at a distance of the order of 10^{18} cm, which means, as we shall see below, that they have only a small effect in ionizing the interstellar gas around the sun.

The main reaction of the solar ions is charge transfer, which takes place with a cross section of approximately $2 \cdot 10^{15}$ cm^2 for protons of 1 keV or 440 km/sec. $5 \cdot 10^{14}$ H atoms/cm^2 are therefore sufficient to convert the proton flux largely into one of fast neutral hydrogen atoms, whereas, instead of the original fast protons, we now have relatively slow ions, i.e. ionized interstellar H atoms. If this happens mainly outside the shock front, the fast or "hot" neutral H atoms will have almost random direction, though the total net outward flux of mass is still given by the constant m (defined previously). The ultimate loss of the ions and electrons must be caused mainly by radiative recombination which, however, requires a much larger volume. For this we note the expression for the lifetime for radiative recombination of hydrogen plasma (see Dessler, ref. I-20).

$$t_r = 3 \cdot 10^9\ T^{3/4}/n^2 \sim p^{3/4}/n^{11/4} \quad .$$

In order to find the volume in which the radiative recombination is effective, we consider that the lifetime, as given by the last expression, must be equal to that defined by the solar production rate, the average density, and the volume. If this is done, we arrive at radii of this volume of the order of 10^3 a.u. (for $T \approx 10^4$ and $n = 10^{-0.5}$) or somewhat more depending on the precise choice of the value of T and n. For the outer parts of this volume, the value of T should be of the order of 10^4 and those of n between 10^{-1} and 1. This is, therefore, the volume in which at least partial ionization is maintained by the solar wind. Between the shock front and the outer boundary of this transition region, there must be extensive mixing of the lines of force of the solar magnetic field with those of the interstellar magnetic field, in which plasma instabilities such as the so-called interchange instability should play an important role.

The formula given before can also be used to prove the point, which was made earlier in this subsection, that the Lc quanta should have little effect on the state of ionization, owing to the large volume involved.

With respect to the particle flux, we observe that this would have a magnitude of only 10^2 particles/cm^2 sec near 10^3 a.u., which means that the diffusion or bulk velocity can be small. The solar energy emitted as kinetic energy of the solar wind results in a flux at 10^3 solar radii of the order of some 10^{-7} erg/cm^2. This flux could be carried without much difficulty by waves of various kinds, since the overall energy density will be of the order of 10^{-12} erg/cc and the velocity of the waves in question would be of the order of, but somewhat larger than, 10^6 cm/sec.

The number of neutral H atoms in the transition region would therefore depend on how they diffuse from outer space through the transition region. Under the circumstances considered here, it is well possible that many of them pass the transition region without being hindered by collisions in their fall towards the sun. We note that at 10^3 a.u. the gravitational escape velocity is of the order of one km/sec, taking into account the radiation pressure in the solar Lyman α line; if such atoms have crossed the transition region and the shock front, they could come relatively close to the sun, their minimum distance being given by the probability of charge transfer or photoionization which is close to 10^{-6} sec (probably $\approx 10^{-5.9}$ for average conditions) at 1 a.u. Atoms falling in from the direction of the apex of the solar motion will have the highest chance to come relatively close to the sun. The ionization probability, as a function of solar distance σr, for an interstellar H atom moving directly towards the sun is given by the integral (α_o is ionization probability at 1 a.u., v is relative velocity, $10^{26.35}$ cm^2 = $(1\text{ a.u.})^2$.

$$\int \alpha_o \frac{(1\text{ a.u.})^2}{r^2}\, dt = \alpha_o \cdot (1\text{ a.u.})^2 \int_\infty^r \frac{dr}{r^2} \frac{1}{v} = \alpha_o \frac{10^{26.35}}{v_\infty} \int_\infty^r dr$$

$$\left(\frac{1}{r^2} - \frac{2 \cdot 10^{13}}{r^3}\right) = \frac{10^{14.5}\text{ cm}}{r} \left(1 - \frac{10^{13}\text{ cm}}{r}\right)$$

in sufficient approximation, since the probability becomes = 1 near $10^{14.1}$ cm. To the extent to which such atoms contribute to the sky background in Lyman α, they should produce an intensity maximum in that direction. Unfortunately, no observations are available so far which could decide whether or not such a maximum exists.

References of Section VI

1) L. Davis, Phys. Rev. 100, 1440 (1955).

2) F. H. Clauser, Johns Hopkins University Laboratory Report AFOSR TN 60-1386, November 1960.

3) E. Parker, Interplanetary Dynamical Processes. Interscience Publishers, New York (1963).

4) W. Axford, A. Dessler, B. Gottlieb, Ap. J. 137, 1268 (1963).

5) T. Patterson, F. Johnson, W. Hanson, Planet. Space Sci. 11, 767 (1963).

6) J. C. Brandt, Icarus 3, 253 (1964).

7) A. Hundhausen, Planet. Space Sci. 16, 783 (1968).

8) A. Dessler, Rev. Geophys. 5, 1 (1967).

9) L. Biermann, B. Brosowski, H. U. Schmidt, Solar Phys. 1, 254 (1967).

10) C. Heiles, Ap. J. Suppl. 15, 97 (1967).

SUBJECT INDEX

(See also: TABLE OF CONTENTS and LISTS OF TABLES AND FIGURES)

SUBJECT INDEX, continued

SUBJECT INDEX, continued

SUBJECT INDEX, continued

SUBJECT INDEX, continued

SUBJECT INDEX, continued

SUBJECT INDEX, continued

SUBJECT INDEX, continued

SUBJECT INDEX, continued

SUBJECT INDEX, continued

SUBJECT INDEX, continued

SUBJECT INDEX, continued

SUBJECT INDEX, continued

SUBJECT INDEX, continued

SUBJECT INDEX, continued

SUBJECT INDEX, continued

SUBJECT INDEX, continued

SUBJECT INDEX, continued

SUBJECT INDEX, continued

SUBJECT INDEX, continued

SUBJECT INDEX, continued

SUBJECT INDEX, continued

SUBJECT INDEX, continued

SUBJECT INDEX, continued

SUBJECT INDEX, continued

SUBJECT INDEX, continued

SUBJECT INDEX, continued

INDEX OF NAMES

(See also: copyright page and LISTS OF TABLES AND FIGURES)

INDEX OF NAMES, continued

INDEX OF NAMES, continued

INDEX OF NAMES, continued

INDEX OF NAMES, continued

INDEX OF NAMES, continued

INDEX OF NAMES, continued